MY MISSION TO CLEAN UP THE WORLD'S MOST LIFE-THREATENING POLLUTION

BY RICHARD FULLER
WITH DAMON DIMARCO

Praise for

THE BROWN AGENDA

"Richard Fuller has been the driving force behind two huge environmental agendas: corporate sustainability and pollution in the developing world. He's the Elon Musk of the environment. An extraordinary man, tackling pollution—one of the world's biggest problems—head on."

—Juan Romeo Nereus Olaivar Acosta
Presidential Advisor for the Environment,
Office of the President, Government of the Philippines

"There are few more important stories than how we humans foul the planet and ourselves, often heedlessly. And there are few better placed to tell that tale than Richard Fuller, who has been working for years to show how we can—and, more importantly, should—clean up the earth. *The Brown Agenda* is the story of an important humanitarian cause revealed in clear and concise prose that is at once deeply necessary and deeply satisfying."

—David Biello
Editor, Environment & Energy, *Scientific American*

"*The Brown Agenda* changes the framework—always the way to have the biggest impact. All of us, but especially those living beyond the significantly cleaned-up West, are daily being environmentally poisoned (not too strong a word). This biggest of all the world's health threats, however, can be readily fixed. Rich Fuller has shown how to do so—from abandoned wastes in Ukraine to old battery factories in Indonesia. This is the rare book that gives the reader the power to act."

—Bill Drayton
Former Assistant Administrator, U.S. EPA
CEO, Ashoka: Innovators for the Public

"Turning 'brown' into 'green' has become a necessary journey for all of us as resource pressures are compounded by the ultimate threat intensifier—climate change. This story, a life's work, shows the difference we can make if we all make it our business."

—Rachel Kyte
Vice President, World Bank

"If you're browsing this book wondering why you don't already know about Rich Fuller and Pure Earth, the answer's obvious by a few pages in: he's been too busy saving lives to 'build the brand.' In fact, Fuller appears only yesterday to have accepted that he even needs a freaking brand. In an age when nonprofits exist to raise awareness and pay themselves, Fuller is out there getting it done. And his memoir isn't some kind of moralizing lecture; it's more like a romp. Who knew cleaning up toxic waste could be the stuff of Indiana Jones-style adventure?"

—Bradford Wieners
Executive Editor, *Bloomberg Businessweek*

“Pollution kills three times more people than malaria, HIV/AIDS, and tuberculosis combined. It’s the single largest health problem we face. Pure Earth/Blacksmith Institute is leading the way toward a better world for all of us.”

—Ben Barber
Journalist, former Senior Writer for USAID

“It takes a particular kind of environmentalist to make his life’s mission ‘brown’ problems in the global south, rather than green ones in the north. Glamorous work it isn’t, but the impact of the author, and those whom he has inspired, on some of the world’s worst pollution problems has been simply incredible. Ideas, energy, passion, stamina, cheek, and brilliant communication skills—Rich Fuller has them all, and his book is an inspiration to us all. This is a captivating story of just how much difference one person can make.”

—Gareth Evans
Foreign Minister of Australia 1988–1996,
President of the International Crisis Group 2000–2009,
Chancellor of the Australian National University

“Rich Fuller has brought evangelical zeal to highlighting the deleterious impacts of environmental pollution and contamination on public health across the world. He has not been satisfied with merely identifying problems but has aggressively sought solutions as well, both through national and international actions. This book is an eloquent testimony to his obsessions for making the world a healthier, safer, and cleaner place, especially for its children. The world needs more Rich Fullers.”

—Jairam Ramesh
Minister for Environment, India, 2009–2011

"Starting with a powerful recount of manmade environmental disasters around the world and their consequent grave dangers to humanity, narrated in Richard Fuller's typical no-nonsense style and without pulling punches, the book also forcefully reflects Richard's passion and commitment to make the world a better place to live and breathe in.

What is refreshing about this book is that it is not a doomsday naysayer. On the contrary, it is about hope and optimism. This is the heroic saga of a heroic man who has devoted his life, his boundless energy and commitment, and most of all, his empathy and his humanity to the cause of 'cleaning up.'

His is a message of determination, of conviction that 'creating a pure earth is a problem for realists,' and evidence that indeed it can be done. A heroic tale indeed!"

—Rajat Nag
Managing Director General, Asian Development Bank, 1988–2013

"Rich Fuller's passion and commitment towards educating people on the under-reported issue of environmental pollution, as well as finding innovative solutions to solve the problems, has helped save the lives of thousands of adults and children around the world. He is a truly inspiring human being and I am honored to know him."

—Dev Patel
Award-winning actor for his performance in *Slumdog Millionaire*

Published by:
Santa Monica Press LLC
P.O. Box 850
Solana Beach, CA 92075
1-800-784-9553
www.santamonicapress.com
books@santamonicapress.com

Printed in the United States

Santa Monica Press books are available at special quantity discounts when purchased in bulk by corporations, organizations, or groups. Please call our Special Sales department at 1-800-784-9553.

ISBN-13 978-1-59580-083-1

Library of Congress Cataloging-in-Publication Data

Fuller, Richard, 1960-
The Brown Agenda : my mission to clean up the world's most life-threatening pollution / by Richard Fuller ; with Damon DiMarco.
pages cm
ISBN 978-1-59580-083-1
1. Hazardous waste sites. 2. Hazardous waste site remediation. 3. Pollution. 4. Fuller, Richard, 1960—Travel. 5. Environmentalists—Australia. I. DiMarco, Damon. II. Title.
TD1052.F85 2015
363.73—dc23

2015010818

Cover and interior design and production by Future Studio
Cover photo by Peter Hosking

For both our families:
Sandy, Alice, Max, Milo, Jessica, and Ethan.

And, of course, for Karti!

CONTENTS

“The problems that exist in the world today cannot be solved by the level of thinking that created them.”

—Albert Einstein

FOREWORD

by Bryan Walsh
Foreign Editor and Senior Writer on Energy and the Environment for *Time* Magazine

In Rudnaya Pristan, I learned a basic fact about environmental science, one that has informed my reporting ever since: nothing goes away. The village of a few thousand people sits on Russia's Pacific coast, over 4,000 miles and several time zones east of Moscow. The summers are hot and muggy, breeding hungry mosquitoes, and the winters are simply brutal. Just about the only reason to live in Rudnaya Pristan was the work provided by the lead smelting plant, which had opened in 1930 and processed the product of the local mines. Many of the bullets used by the Red Army to defeat the Nazis came from the Rudnaya Pristan smelter, which operated for decades.

By the time I had arrived in the village on a reporting trip in the summer of 2007, the smelter had long since closed. But its environmental legacy was still felt. Lead had contaminated the very soil of the village, the crops its people ate, and the fields its children played in. Blood lead levels among children in Rudnaya Pristan and the neighboring town of Dalnegorsk were as much as eight to twenty times above maximum allowable levels in the U.S. Lead is a potent neurotoxin, and in high levels it can damage the brain and the kidneys. It's especially dangerous for young children, whose developing brains can be permanently damaged by lead contamination. The youth of Rudnaya Pristan hadn't even been born when the smelter closed down, but they were paying the price.

Lead contamination isn't the sort of environmental problem

that gets a lot of attention. It's invisible, it's not connected to climate change, and it has nothing to do with endangered animals. But lead contamination, and other examples of legacy industrial pollution, exacts a tremendous cost on human health—especially on the poorest of us. In a 2013 study, researchers looked at nearly 400 toxic sites in India, the Philippines, and Indonesia, and found that elevated levels of dangerous chemicals like lead and chromium have a direct impact on mortality. The number of disability-adjusted life years lost because of toxic wastes in those countries was over 800,000, worse than the morbidity and mortality caused by malaria.

And lead is just the start. Radioactivity from nuclear plants in Chernobyl, toxic chemicals from electronic waste (e-waste) in Ghana, mercury gold mining in Indonesia, crude oil spilled in Nigeria . . . While the developed world has mostly cleaned up the worst of its industrial pollution, much of the developing world is still being poisoned on a massive scale. According to one analysis by the Global Alliance on Health and Pollution, air and water pollution and toxic waste killed 8.4 million people in 2012—nearly three times the number of people killed that year by malaria, HIV, and tuberculosis combined. And these catastrophes are unfolding under the radar, in tiny towns like Rudnaya Pristan, far from the attention of most of the world.

But not from everyone. In 1999, Richard Fuller started the Blacksmith Institute, now called Pure Earth, to address the world's forgotten environmental challenge. I was in Rudnaya Pristan in 2007 to see Pure Earth at work, taking samples of the contaminated soil and working on a remediation program. Because the smelter was no longer operating, the solution was simple: remove the contaminated dirt, starting in the places where children gathered, like parks and school fields.

The good news about industrial pollution is that it doesn't take international treaties or technological miracles to solve. It just requires will and work—two things that Fuller has exemplified in his career at Pure Earth. And the payback can be enormous. A 2007 study done by Pure Earth found that, for every $1 to $50 spent, pollution cleanup projects gained back a year of human life. Compare that to $250 to $300 for every year of life gained by projects that focus on infectious disease or malaria.

That isn't to argue that we shouldn't be spending money on TB and HIV solutions, or on projects to avert climate change. Far from it. But those challenges already have very high-profile champions who spend billions of dollars a year. Fuller has taken on industrial pollution essentially by himself—and that makes the success that he and his organization have achieved all the more impressive. Add up the numbers, and you'll find that Pure Earth has cleaned up scores of polluted sites in countries around the world, affecting millions of lives for the better. And Fuller has done this with a fraction of the budget enjoyed by larger environmental groups.

But there's so much more to do. This isn't glamorous environmentalism; there are no poster-worthy polar bears or ice caps involved. Fuller's work is carried out in slums and broken villages like Rudnaya Pristan, where he is demonstrating how we can save millions of the world's poorest people from long-term health problems and early death. This isn't about saving the planet; it's about saving ourselves from ourselves. We just have to open our eyes—and pay attention.

NOTE TO THE READER

Some names and locations have been changed to protect the privacy of the actual individuals.

INTRODUCTION

It started out simply enough: peering into an open-cut mine and watching my feet to make sure I didn't slip on the ice. The pit was several hundred feet deep and about half a mile across. We were at the edge of an old cinnabar mine near the town of Horlivka, in the eastern part of Ukraine.

The Russians had mined the cinnabar for mercury but the pit had lain abandoned for some time. I was there with a dozen other environmentalists in a training workshop to identify polluted sites. This place certainly qualified. High levels of mercury had been found in the soil of a nearby village; we were there to assess the risk. My biggest concern at the moment, however, was keeping myself from slipping on the icy edge of the mine and falling in. One glance was enough to tell me it was a long way down, and dark.

Vladimir was our local coordinator. He tugged at my elbow.

"Uh-oh," he said in a low voice.

I followed his gaze and saw two big black SUVs pull off the road about fifty yards away. A window in the lead vehicle rolled down, and someone within barked a few words in Russian. Vladimir scurried over, listened for a moment, and jogged back. The look on his face was grim.

"They want you to go with them," he said.

"Who?" I asked. "What do you mean? Who are they?"

"I'm not sure, but they are officials. We have to do what they say."

Things unsaid lingered in the air as we both walked over, and Vladimir introduced me.

"This is Richard Fuller, the leader of our investigation team." He said this in Russian, then repeated it for me in English.

"Get in," someone said. In English. A back door to the SUV opened and I caught a glimpse of two heavyset, grim-faced men inside.

"You too, Valodya," I said, hoping my use of Vladimir's colloquial name would break the tension. It did not. Vladimir turned back to our training group, which was struggling toward us en masse, their faces curious and strained. He gave them some instructions, then opened the front door and got in the front seat. We pulled away with the second SUV following close behind. I craned my neck and felt some relief when I noticed the white van our team had rented filling rapidly. A moment later, it was following us.

At least there are witnesses, I thought.

I was distracted by the sound of Vladimir having a detailed and heated discussion with a man I assumed was the boss, the older gentleman sitting next to me. I did not dare interrupt. If my group was doing something illegal—or even if these men had simply decided we were doing something illegal—my safety now depended on Vladimir's gift for talking his way out of difficult places. This was not the first time we'd needed that skill, but it seemed by far the most serious. From everything I could gather, these guys who had picked me up were not drunk soldiers or low-level officials looking for a bribe. They were clearly big-time operators of some sort. In the former Soviet Union, that can be deadly.

The conversation went on for some time, until Vladimir glanced at me. He must have read the look on my face.

"Rich, is okay," he said. "Everything is fine. This is the mayor of the town—Horlivka, over the rise. He has not come because we are in trouble."

"Great," I sighed. "Then what's this about?"

"He says they need our help."

In 2009, the city of Horlivka, Ukraine, was literally a bomb waiting to go off. The ruins of a secret former Soviet weapons factory sat on a 400-acre campus in the middle of town. Within its crumbling concrete halls, stacks of corroding metal barrels leaked deadly mononitrochlorobenzene, or MNCB. The Soviets left this cache behind when their empire collapsed in 1991. It was very bad news for the people of Ukraine. MNCB is so toxic that just half a teaspoon ingested, inhaled, or absorbed through the skin can kill a human being. The barrels at Horlivka held over 8,000 tons of the stuff, enough of the toxin to wipe out every living thing in a radius of many miles. You could often smell the stuff in the air of the town, a faint, sweet almond odor.

Along with the MNCB, a flammable toxic compound called trinitrotoluene—more commonly known as TNT—was left lying around in pipes and underground sarcophagi. TNT was one of the factory's end products. When the Soviets abandoned the factory, they left nearly ten tons of it lying around, much of it in a form that was volatile, ready to explode. Other buildings nearby contained large amounts of various highly corrosive acids.

The mayor and his team drove us up to the entrance of this facility and pointed to the buildings. Night was falling. They made no move to get out of the car, just kept talking to Vladimir in Russian. Curious, I opened my door and got out. The other members of my team had pulled up behind me in our white van. They, too, got out and came over to join me. Once they'd established that no one was being arrested, a few of them drifted over to talk with the mayor and his people. But no one made a move toward the plant. We all just stood there, watching it from a safe distance.

At the time, about 260,000 people lived in Horlivka. Chemicals from the plant had leaked into the air they breathed and the water they drank. Local doctors had pegged the average life expectancy in Horlivka at under fifty years old. But there was something even more worrisome. Up until very recently, the people of Horlivka lived under the constant threat of the place exploding.

It wouldn't take much—just a stray bolt of lightning, a cigarette butt tossed aside, or a spark borne aloft on a hot dry wind. The slightest incendiary provocation could have set off an explosive chain reaction. The TNT would go up in a flash, tossing toxins throughout the town. According to some estimates, a disaster at Horlivka could have easily dwarfed those experienced at Chernobyl and Bhopal combined. Experts have called Horlivka one of the worst instances of legacy contamination in human history, a deadly threat to both the environment and human health. Yet practically no one has heard of it.

This was the most dangerously polluted place I had ever seen, worse than the mercury mine by a mile. Eventually, my team and I dispersed after promising we would do our best to help the mayor and his people with their toxic legacy.

A month later, we sent in a team that included the top environmental scientist from the U.S. Army (now retired). Ira May took a few steps into the first building at the abandoned weapons plant. He read the names marked on some leaking bags of chemicals just inside the door and his face went white.

"Everyone out. Now." His voice was low. Insistent.

Everyone obeyed him promptly.

Back outside, he told everyone present: "No one goes in there without full protective gear. That's nasty stuff in there. Does everyone understand me? Good. Then let's suit up."

That's how our project to deal with Horlivka and its toxic legacy began. Before I tell the whole story, it's important that you understand how pollution of this sort—what I and a lot of other people have taken to calling "the brown agenda"—isn't confined to one problem in one city of one country. It's a global issue, it's spreading fast, and we have to do something about it *right now.*

For instance, a few thousand miles away, in Delhi, India, the air is getting worse. Particulates from low-grade diesel exhaust combine with coal exhaust from a couple of nearby power plants to create a haze that descends over everything, rendering permanent twilight on bright, clear days. The World Health Organization has set the level for safe, breathable air at twenty micrograms per cubic meter. During one of my visits in June and July of 2014, the air in parts of Delhi registered particulates at 600 micrograms per cubic meter on at least five separate occasions. A recent analysis in the *New York Times* found the air in Delhi to be twice as polluted as that in Beijing, the poster child of bad air.

Conditions like these have taken a toll on the citizenry, especially in the poorer districts. Living in Delhi means living with a constant cough and shortness of breath. Incidents of chest infection, pneumonia, and asthma are rampant. Expats I know have cut short their three-year assignments after watching their children suffer, unable to exercise or play outside. Most Indian children will never get to leave, of course. Delhi is their home and their sentence in one.

Travel farther east, to rural parts of Central Kalimantan, Indonesia. Small-scale gold miners rip apart the Kahayan River with sluicing hoses, searching for gold-rich rocks. In the forests all around them, birds and an occasional orangutan watch in dismay as this paradise slowly becomes lost. The miners process the ore by hand, using a primitive technique that involves liberal quantities of toxic liquid mercury. They don't take care to clean up after themselves, so each year, tons of mercury enter the local environment. The element poisons the miners and their children, many of whom become stricken by tremors. (In Victorian England, this was known as Mad Hatter's Disease, since haberdashers used mercury to stiffen the felt of the hats they were shaping.)

But the damage doesn't stop there. Apart from being a toxic heavy metal, mercury is an element, and therefore persistent—it does not break down into harmless, smaller particles. Once it infiltrates the river and local marine life, it sweeps out into the ocean, where it joins the global food supply. Big game fish like tuna can become especially polluted by it. These days, everyone in New York, Los Angeles, Tokyo, and other major cities has heard the warnings. *Caution: Mercury Levels Rising in Fish! Avoid Tuna, Especially When Pregnant!* What most people don't know is where that mercury comes from. A great deal of it leaks from coal-fired power plants around the globe. But the largest amount originates from impoverished miners who make their living panning for gold in the ancient jungles of Borneo and other places.

Not far away, in metro Jakarta, a half-dozen urchins play soccer in a local field using a ball they've fashioned from wadded cloth and duct tape. Filthy hand-me-down clothes hang off their spindly frames. The soil beneath their brown bare feet glints light grey in the sunlight of the tropical afternoon; it smells faintly metallic. The children have no idea this field once played host to an unregulated car-battery recycler. They don't know the recyclers buried their waste and fled when officials threatened to shut them down. Nor can they comprehend that the land they are playing on is poisoned by lead dioxide, a chemical so toxic it causes brain damage and nerve deterioration before it kills.

Readings taken at the soccer field in Jakarta with a handheld laser spectrometer show lead levels measuring 49,239 parts per million of lead, a quantity that massively exceeds the standard of 400 set by the World Health Organization. Arsenic readings are also off the charts at 1,744 parts per million; the common approved standard is less than ten. Meanwhile, the kids dream that football will get them out this slum. They don't know how dangerous it is. They're focused on scoring a goal for their team.

These contaminated sites are everywhere. Unsafe recycling in backyards accounts for one-half of all cases of car battery recycling in the developing world. This industry affords a basic standard of living for a few while poisoning the communities around them irreparably. Lead is responsible for killing five of Seynabou Mbengue's ten children. Like many families in Ngagne Diaw, Senegal, Seynabou used to make her meager living recycling old car batteries. She would break the battery casings by hand and dump the acid out on the ground before extracting the valuable lead components within and melting them over an open fire. She had no idea how dangerous this was until her last five babies suffered convulsions and died, all before the age of two.

"That's when I made the connection," she said. "[While] pregnant or breastfeeding my babies, I never stopped recycling batteries. I am still full of lead."

Seynabou's story illustrates how much more vulnerable babies and children are to pollutants. They die more often and faster than the adults around them, whose larger bodies can withstand the toxic chemical load much longer. Along with Seynabou's five toddlers, twenty-seven more children are known to have died from lead poisoning in Ngagne Diaw. The true toll is likely much higher.

In a modern world so often focused on "going green," these are just a few troubling examples of what we call "brown" issues: pollution gone awry. In 2012, pollution like the types I've just described killed 8.9 million people worldwide. To put this into perspective, WHO statistics showed that fifty-five million people died in 2012 overall—that's every person who passed away on the planet, whether it was from car accidents, suicides, old age, cancer, hospital errors, being struck by lightning, infectious diseases, parachute failures, war, or what have you. In other words, in 2012, brown problems—pollution—killed about one in seven people.

The breakdown of these 8.9 million deaths falls like this: 3.7 million died from contaminated outdoor air; 4.2 million died from exposure to particulates in indoor air from cooking stoves; about one million died from chemicals and contaminated soil and water; and 840,000 died from poor sanitation. All of this data comes directly from WHO's website and databases, except for the soil statistics, which I sourced from more recent numbers generated by the Global Alliance for Health and Pollution, and these numbers are likely understated. (For those of you paying attention to the math, some air pollution deaths overlap in both indoor air and outdoor air. Hence the lower than expected total.)

These deaths, which are additional to the "normal" levels of death, would be avoided if pollution were not present.

In the same year, 625,000 people died from malaria, 1.5 million died from HIV/AIDS, and 930,000 died from tuberculosis. As you might know, this trio of terrible diseases draws over $20 billion per year from international charities and governments. However, HIV, malaria, and TB combined only kill one-third the number of people that pollution does.

Here's another statistic worth pondering: of the 8.9 million people killed by pollution in 2012, 8.4 million resided in low- and middle-income countries. In other words, the brown agenda is not a "rich country" problem. Overwhelmingly, it's a problem confined to the developing world.

And one more thing. It's important to note that pollution rarely kills people directly or quickly. Instead, it causes heart disease, chest infections, cancers, respiratory diseases, or diarrhea. Pollution acts as a catalyst, increasing the rates of these diseases above those by which they would normally occur. Scientists make these estimates from studies that measure disease in places with and without pollution. The studies go through a detailed peer-review process to check their calculations.

Which is why WHO considers pollution a "risk factor"—a

similar threat to human health as obesity, smoking, malnutrition, or poor exercise. But pollution is the king of all risk factors. Worldwide, its fatality numbers dwarf those caused by any other risk factor in any other context whatsoever.

So what's being done to combat it?

Sadly, very little. As I've already mentioned, HIV/AIDS, malaria, and TB attract lots of funding, as does biodiversity, and, to an even greater extent, climate change. Don't get me wrong, these are all important issues and worthy of our contributions. But international aid dollars to combat pollution are almost nonexistent—perhaps $100 million a year at best.

In all fairness, the situation has improved somewhat in the last few years. The governments of countries like China and India have realized that pollution is a threat to their ongoing economic development and have started allocating funds, though only small amounts so far. Meanwhile, millions of people keep dying each year—children foremost among them—with many more likely to come.

So why aren't we in the wealthy countries doing anything to stop this?

Part of the reason is lack of focus. Part of it is lack of awareness.

Backtrack to the 1950s and '60s. In the early days of the United States' environmental movement, pollution and biodiversity were seen as two intertwined focuses. Rachel Carson's harrowing book, *Silent Spring*, brought the problems of pesticides and environmental degradation to our doorsteps. A bit later on, toxic calamities such as the ones found at Love Canal in New York and the Cuyahoga River fire in Ohio galvanized the public to demand laws protecting our soil, air, and water.

The EPA was formed in this era and, without question, our country is better off under this agency's stewardship. Regulations set limits for air, water, and soil pollution while establishing that non-compliance can bring claims for compensation. If you live in a toxic place, lawyers can help. The big companies have become more responsible, worried not only about their workers, but also about their reputation and stock price. At present, there are dozens of terrific nonprofit groups watchdogging polluters.

For all these reasons and more, real instances of unaddressed, harrowing pollution are few and far between in rich countries. Once in a while, some dramatic event will make the news cycle. A dam has burst somewhere, contaminating local water supplies. Or there's an inexplicable rise in autism, quite likely caused by the enormous number of chemicals we ingest on a daily basis. Lately, we've heard how endocrine disrupters are on the rise and can potentially affect our reproductive capabilities.

But terrible air? Contaminated drinking water? Poisoned soil in our schoolyards? Wealthy countries don't really suffer these things. Conditions like these are someone else's problem. Remember that WHO statistic I mentioned? In 2012, ninety-five percent of deaths caused by pollution took place in low- and middle-income countries. Read: we in the West enjoy clean water and air; it's the rest of the world that's messed up. We enjoy the luxury of focusing on going green, whereas other countries the world over must suffer their way through conditions we can only call brown.

In poor countries, there are thousands of places where manmade toxic pollutants have taken root and spread like cancer. You will find them in all low-to-middle-income countries such as India, China, Russia, Peru, Azerbaijan, Indonesia, Ukraine, Mexico, Zambia, and the Philippines. In almost every instance, the culprits responsible for these contaminated sites are the governments, businesses, and people that race to increase their industrial capacity at the expense of their citizens and environment.

Interestingly, the worst polluters tend to be local companies, rather than multinationals. By most comparisons, big Fortune

500 companies run their operations well. They don't buy into dirty industries because they know they need to uphold Western standards, at very least to appease Western shareholders. The local companies and small backyard industries do the bulk of the damage. Small-time operators running backyard businesses can do immeasurable damage to their local populations and environment. They can ruin whole tracts of land for generations to come, to say nothing of human beings.

It's hard to imagine just how bad it can be, so let me present a scenario.

You wake up each day on the dirt floor of the shack you and your family lashed together with cast-off materials from the nearby construction site for a five-star hotel. Your last baby was stillborn blue, probably from heavy metals poisoning. Your husband works seventy hours a week sorting chemicals in a badly run pesticides factory. Lately, he's come home coughing up blood. He looks thinner and more exhausted each week, and you want to tell him to stop, but how can you? The pennies he earns are the only things feeding your kids.

So you head to the local pond with your plastic bucket. The water you scoop from the pond is brown and stinks of human waste, but there's nothing else you can drink. You tried straining it through cheesecloth, but it doesn't do much good. Meanwhile, the factory next door to your slum, the one the government came and shut down a while back, has started operating again, but only at night. Its chimneys pump out serpents of thick smoke that pulsate in the dark, their tails lit bright by the flames of a forge, and God only knows what's burning. Last week, your eldest child started coughing through the night. The rest of your children are sickly and slow to learn even the most basic concepts. As years have passed, you have watched them grow duller and more enfeebled. And none of your friends or family can help you since, curiously, almost everyone in your neighborhood has the same problems.

You are one of the poisoned poor, without a voice and without any hope. Regulations that might exist to combat the conditions are never enforced. You cannot simply pick up and move to another town. It took you years to establish yourself to this extent, and anyway where exactly would you go? Every village shares the same plight. Like the rest of the world's underprivileged, you have become cannon fodder in the ongoing war of growth.

There are bright points in all this. Take the case of Horlivka. The city faced grave peril, but it could be saved, and save it we did (more on that later). For now, just know that the worst toxins jeopardizing Horlivka have been removed. There are other success stories like this, stories that show we can prevail against brown conditions if only we marshal the will.

In the last decade especially, overall levels of sanitation and clean water have shown dramatic improvement. Millions of people now have safer water supplies, thanks to great programs by governments like that of China and organizations like Charity: Water. That's a huge improvement. Ten years ago, 1.6 million people died each year from improper sanitation. This figure is now down to 840,000. Moreover, some cities, like Mexico City, have worked up a means to control air pollution. Air quality there has really improved, though on some days it is still quite awful.

The bottom line is that what we're doing is clearly not enough. I realize that you might disagree. Perhaps you're one of those people who thinks, "Why should we care? I mean, what you're saying sounds horrible, but it's taking place in some foreign country. That's not our problem, right?"

Wrong. Brown problems affect us both directly and indirectly. Some of the toxins I've mentioned—mercury serves as a prime example—end up in our supermarkets. More subtly, however, pollution kills one in seven people, many of whom are engaged in the mining and manufacture of products we take for granted, products necessary for our comfort. I would argue that we owe these people a debt of responsibility, though we also need to deal with this problem for own security.

Consider a study conducted by Denmark's Ministry of Foreign Affairs in 2000. It concluded that environmental burdens have actually grown in recent years, at least in part because political and economic bickering has allowed polluted sites to proliferate.

By tackling cases of toxic pollution the world over, we score big wins for both the environment and humanity. And doing so is relatively simple. The technology to remediate even the worst toxic sites exists within the industrialized world, and remarkable cleanups are often achieved with shockingly small sums of money. As I mentioned, in order to succeed, we just have to rally our collective will. We just have to see these hellholes as our common enemy. When we prioritize fixing them, everybody wins.

I want to show you how, by working together, we can score big wins for our planet and our species for generations to come. We can help to create a newer and better world, a pure earth free from the ravages of brown. But before I do this, I want to tell you how my journey began.

—**RICHARD FULLER**

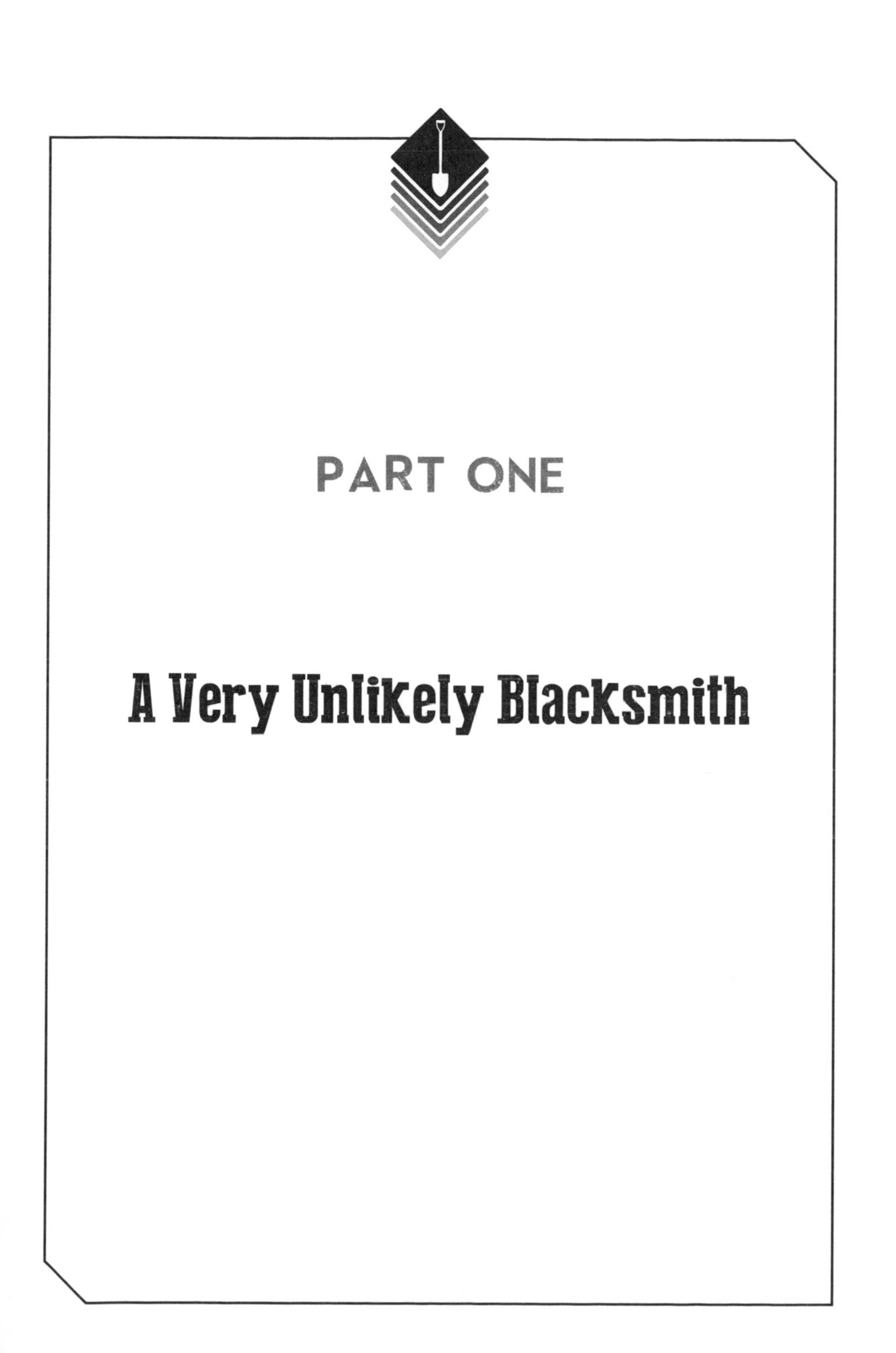

PART ONE

A Very Unlikely Blacksmith

CHAPTER ONE

Two Roads Diverged in a Wood

One of the most important lessons I've learned is that you can't change the world by punching a clock.

In the late 1970s, I was a student at the University of Melbourne, training to be an electrical engineer. Like many college students, I found myself searching for ways to support a comfortable lifestyle. Back in those days, all the electricity in my region was generated and controlled by a huge government agency called the State Electricity Commission of Victoria. This commission ran a scholarship program that paid college students to study, so long as they promised to work for the agency once they graduated. This struck me as the perfect combination: I would get paid to go to school and was guaranteed a good job once I got out. I signed up for this program without any hesitation. As a result, during my last two years of university, I was the Guy Who Had All the Money, which did wonders for my social life. But that's another story for another time.

Once I'd earned my bachelor's degree, I went to work at the SECV. I was tasked with designing power lines for new subdivisions. It was not particularly demanding work, and I got bored quickly. Developers had to apply to the SECV and ask us to run power out to their new housing sites. Someone like me had to figure out all the essentials. How many poles would we need to run and where? How far apart should we space them, and how tall should they be to allow for low-hanging wires in the extreme heat of summer?

If any of this sounds like a complicated engineering exercise, I assure you, it wasn't. We had this process down to a science. I ran each application I received through a model, drew up an installation plan for the power grid, and then filled out a cost sheet that charged the developer for all the work that would have to be done. In any given week, four or five of these jobs might land on my desk. It took me a few weeks to master the system, but once I did, I could knock out a whole week's application load in a matter of hours, perhaps even in a single workday. Which meant I had the remaining four days in the workweek to do . . . well, nothing.

I went to my boss and told him I needed more work. He assigned me a few more projects to do. Now, instead of working six or seven hours per week, I was working ten, sometimes twelve. Once I'd knocked those projects out, I would wander the floor, asking my colleagues if they had anything I could help them with, but they never did. Whenever I approached their desks, they would hunch over their quite small workloads and pretend they were locked in arduous labor.

Or maybe that's not being fair. I mean, maybe for them, their workload really was arduous labor. Honestly, this was my first time working in a bureaucracy. I hadn't imagined how clunky that overall mechanism could be. But from what I saw, people got rewarded for finding new ways to do nothing rather than doing their jobs well, or making the organization more effective.

At any rate, having nothing better to do with my time, I decided to teach myself some new computer languages. This went off quite swimmingly until, one day, my boss swung by my desk and saw all the Fortranplus books lying open in front of me.

"Hey," he said. "You shouldn't be reading those. You should be working."

"Right," I said. "But I've done all my work. See there? My tray is empty."

He wasn't pleased, so I put the books away and donned a pair of headphones instead. For some reason, that office didn't condone extracurricular reading, but headphones were considered perfectly acceptable accoutrements. Everyone wore them. But while other people were listening to music, I started learning German from a set of language tapes I'd bought. Why German? No particular reason. It was something to do when I had nothing to do, and I've always hated having nothing to do.

Some days, I would punch in, go to my desk, listen to language tapes for a couple of hours, and then take myself shopping. I strolled around Melbourne making friends with the clowns and musicians busking on the city's main strip. After that, I'd go back to the office, finish up any small matters that had landed on my desk, and punch out.

I'll say it again: you can't change the world by punching a clock. Back then, I didn't know I wanted to change the world, I just knew I was unspeakably bored, and all at once, the plight I found myself in made a wicked sort of sense. The SECV had paid me so well and for so long so it could shackle me to a desk for the next couple of years and make me do a job that nobody else in their right mind wanted to do.

Eight months rolled by like that. With each passing day, I grew more and more fed up, and less and less willing to hide it. My boss began sending me out to Yallourn, an industrial town about 100 kilometers east of Melbourne, in the Latrobe River Valley.

Yallourn was where Melbourne's power was generated. Four separate coal-burning facilities worked constantly to supply most of the juice flowing through southeastern Australia. I took day trips out there three or four times. My boss told me to review operations, whatever that meant. It was a bullshit assignment. The tours I was always given were hasty at best, almost as if no one had been told I was coming. I began to get the feeling that my boss didn't care what I saw or did in Yallourn, so long as I was out of the office and, not coincidentally, out of his hair.

Each time I got back from Yallourn, I wrote a report on what I'd seen and submitted it promptly, but I never got a response. Each time I followed up, I was told that my recommendations were being considered. One day I found one of my reports in a wastebasket. Clean pages, no dog-ears. The damn thing hadn't been so much as touched.

Back at my desk, I wrote a letter to the chairman of the SECV. I stated my name, position, and department, then set out a few ideas I thought would make the organization more efficient.

"Mostly," I wrote, "I feel as though this commission lacks spirit and drive. I'd like to meet with you sometime, at your convenience of course, to suggest some ideas. Yours very truly, Rich Fuller."

I was twenty-one or twenty-two years old when I sent that letter. A few days later, the chairman's secretary called my office line to set up a meeting. When the day rolled around, I left my cramped, workaday pen, walked three blocks to the commission headquarters, and took the elevator to the chairman's office on the twenty-fifth floor. They didn't leave me waiting long. The chairman came out personally and invited me into his office.

He was such a lovely man. A delight, really. He had the white hair, calm smile, and excellent suit of a political appointee with immense job security. We talked about sports and family, and hit it off right away. His desk phone rang several times, but he picked up the receiver and kept saying, no, no, thank you, cancel that. Then hanging up and waving his hand like, gosh, can't they leave us alone?

I got the impression that he enjoyed talking to someone who was young and enthusiastic and honestly wanted to improve the way the commission was run. The agency was such a staid place. He counseled me to apply for jobs in a new department they were bringing online, and specifically one that had to do with renewable technologies.

"That's the future," he said. "It might not seem like it now, but it is."

I thanked the chairman for his advice and walked back to my office, where my boss was waiting for me, furious. Evidently, someone had tipped him off to where I was. He hauled me into his office, slammed the door, and started to shout. Looking back, I guess I can see why he was angry. The chairman was my boss's boss's boss's boss's boss. By contacting him, I hadn't gone outside the chain of command so much as I'd entered a different universe altogether.

"If you ever . . . *ever* pull a stunt like that again!" my boss yelled.

And on and on like that.

After he'd shouted himself hoarse, he got really quiet and stared out the window while I stood there. It was awkward.

"Tell me what you talked about," he said finally. He sounded like a guilty schoolboy who'd just been tattled on to the headmaster. "Tell me what you said about me."

That was the moment I realized that everything I'd said about the commission was spot-on. The place was a dead zone, a graveyard for creativity. The only people who stayed there were the ones who'd decided to kill themselves slowly, heads hung, punching a clock day in and day out until death showed up at the end of their lives and put them in a box.

After that meeting, I went back to my desk and started looking for a new job.

A couple weeks later, I requested a meeting with my boss. He accepted, but didn't look pleased by the thought. We sat in his office again. This time, I told him I'd been offered a position at IBM, which was true, but there was a problem. According to my scholarship deal, I owed the commission two or three more years of work.

"Great!" my boss said. "No worries, mate! We'll sign out the rest of your contract, sure! Good luck. Great knowing you. Bon voyage!"

They were thrilled to have me out of their hair, and I was thrilled to leave.

At IBM, I enrolled in a general intake program. The company had hired a fresh crop of twenty or thirty self-starters whom they wanted to train and test with an eye toward shuttling each person into one of two career tracks. Salespeople would hit the road, meet prospective clients, and sell different products—computers, software, phone systems, and so on. Engineers would stay in the office and work out any technical details; they were considered support personnel for the sales force.

The company wanted me to become an engineer—my electrical engineering degree seemed perfect for the job—but I lobbied to be made a salesperson. For one thing, being in the sales force meant you were higher up on the company food chain. You met and chatted with prospective clients over bottles of wine in good restaurants. That sounded great to me. Also, I felt burned out on technical matters. The thought of working another repetitive computer simulation was enough to make me run screaming into the wilderness.

To be a salesman, however, you had to pass sales school. The company flew me up to Sydney, where I trained for two weeks straight before being evaluated by the sales force. When I flew back to Melbourne, my boss called me into his office and sat me down.

"I just received a phone call," he said. "Congratulations are in order. You're the first person in the history of IBM Australia to fail sales school."

He hauled out a packet of fax sheets that contained my written evaluations. To this day, I remember the litany of critiques.

Too aggressive. Pitched products in overly technical language. Doesn't grasp how business works. Poor people skills. That last one cut pretty deep.

A silence settled between us.

"So what does this mean?" I said.

He frowned. "The blokes in Sydney said you should stay on as an engineer."

"But I don't want to be an engineer," I said.

My boss nodded and said not to worry, he'd think of something.

As luck would have it, he landed in a new position a few days later. The company made him a sales manager in a brand new division that sold telephone systems.

Have you ever worked for a big firm and dialed "9" to get an outside number? There's a reason why you have to do that. All the company's extensions run to a big computer in the basement called a PBX. Back in the early 1980s, this technology was so new at IBM that no established sales rep wanted to touch it. Why go through the hassle of re-training to sell phones when they were already making fat profits selling computers?

As the new kid on the block, my boss needed people to fill out his sales force. Quality didn't matter so much. First and foremost, he needed warm bodies. So he cobbled together this ragtag team and, when everything was said and done, he had one slot left, which he pitched to me like this:

"Look, we've worked together a bit, haven't we? We get along, right? You didn't graduate sales school, I know. But technically you did *finish* sales school, right? Did all the coursework . . . that's something, isn't it? So, what do you say?"

I said I was sold. I took that job and never looked back.

Within a few months, I started selling phone systems hand over fist. My biggest clients were the Mars Corporation, Australia, which sold candy, Uncle Ben's Rice, and a dog food company, but I also sold to the state government, libraries, and school systems. My commissions blasted straight through the roof. By the end of that year, I was one of the highest-paid first-year salesmen at IBM Melbourne. The year after that, I doubled my book. I could hardly believe my good fortune.

Don't get me wrong; the blokes at sales school had been spot-on when they assessed my flaws. But having had them, I was determined to fix them. I knew I was working against the odds, and knew that I wanted to be a better salesman. So every day I made it a point to learn something new about my job or about business. I stopped concentrating on numbers and focused on building relationships instead. It became my goal to leave every meeting having just made a brand new friend. This, in turn, began to affect my sales numbers. At the end of my second year, I earned a key company sales award that let me travel around the world on the company's behalf.

Not too long after that, I found myself in Manhattan on a large dinner-and-dancing yacht, which the company had rented for an evening. I remember standing at the rail, looking out across the Hudson at the New York City skyline. The skyscrapers rose like the walls of a starlit cliff. They had energy and majesty, and the closeness of them swept me away. It felt like, if I wanted to, I could reach right out and touch them.

This is where I want to live, I thought. Manhattan. The center of so much energy.

But I didn't heed that call to adventure. Not yet. Instead, I went back to Australia, where I had another banner year. My business was earning such fantastic profits that the top brass called me aside and told me they wanted to enroll me in a fast-track management program. The commitment would involve spending a year at the IBM headquarters in Westchester County, New York, then another year in some other country, possibly Germany, maybe Japan. If I did well after a couple of such rotations, the firm would bring me back home, where I would assume a senior management role at IBM Australia.

On paper, this opportunity looked like a dream come true. I would climb the corporate ladder fast. I would receive money, prestige, and perks. In no time flat, I'd be sitting on top of the world. But something about the whole deal made me uneasy and, looking back, I blame the trees.

In my second year as a salesman, two friends and I had founded an environmental nonprofit. The whole thing got started as sort of a lark. We all had stable jobs. We all rented suburban houses or apartments, wore blue suits to work, carried the requisite leather briefcase, and drove company cars. But we all agreed there was something missing in our lives, like we should have been making some kind of difference in the world, and we weren't.

One night, over a couple of beers, we watched a news clip from Los Angeles. The city had just led a campaign to plant a million trees. One million trees! My friends and I thought this was brilliant. Trees are healthy. Trees are alive! Why not do the same thing in Melbourne? That's how we started the Tree Project.

It was a true grassroots operation. We enlisted the aid of Michael Leunig, the famous Australian cartoonist, to work up a Tree Project logo, which we had stenciled on some t-shirts. Then we lined up a few small donors, which meant that we weren't just using our own money—we had people helping us, and therefore we felt responsible to those people. Emboldened by this, we approached the city government of Melbourne in a series of meetings I still remember with delight.

"Why aren't you planting any trees?" we asked the government officials. "You promised that you would."

"Well," they said. "We have so many other things that need to be done. Besides . . . we're planning to."

"But this is such an easy thing to fix, why not just get on it?"

"Manpower," they said. "Resources. Studies. Planning codes." Et cetera. Their list of reasons went on and on.

"Right," we said. "Tell you what. Seems like you really *do* have a lot going on, so why not let us take this off your plate? Give us some money, we'll do the job for you."

When they balked at our offer, we told them how disappointed we were.

"Look," we said, "the thing is, we're going to do this anyway. You can help us or not, that's up to you. But you're going to look really stupid if three friends just out of college get this job done when you can't."

Eventually, they buckled and gave us some funds, and thank goodness they did. At that point, we needed every penny we could lay our hands on.

Planting trees is no simple thing. We had to hire experts, who set us straight on certain matters of higher dendrology. They told us, for instance, that we couldn't just plant native trees, we had to plant native *local* trees. There's a difference, they said. (This was all news to us.) The seeds we had hoped to plant in Dandenong couldn't be used there because they'd come from Geelong and Warrnambool rather than Dandenong itself. I wish I could tell you I understood the reasoning behind all this, but it's out of my expertise. Something to do with insects and root systems, I think. Taxonomy, that sort of thing. The upshot was that we followed the advice we were given. I believe then, as I do today, that planning a project is useful, important work, and a good planner works off the input of experts whose opinions can be cross-matched against one another to generate best practices.

Our grassroots operations started to thrive. I recall one big event the Tree Project held to commemorate planting 10,000 trees in Yarra Bend Park, one of the largest plots of natural bushland left in inner Melbourne. I can't describe how exciting that felt. It seemed right to me in a way that few things have ever felt right to me, one of those moments in your life when you know you're in the right place at the right time, doing the right things for all the right reasons. Whereas the offer I got from IBM felt wrong in every way.

At first, I had a hard time figuring out what was troubling me. The company's offer had so many positive aspects, it seemed crazy that I should find it so upsetting. Hoping to work out my feelings, I went home, opened a bottle of wine, and sat at my kitchen table with a pen and a blank piece of paper. As an exercise, I thought about what I wanted to do with my life and wrote down whatever popped into my mind.

The first thing I wrote down was:

I want to make a difference at a global level.

I stared at the paper for a moment and frowned. This was the first time I could recall thinking such a thing, let alone setting it down in my own handwriting.

Interesting, I thought, and kept writing.

I want to live in a foreign country, learn a different language, and experience a different culture.

I want to find out what's going on with the Amazon rain forests.

Again, I stopped writing and stared at my page. The Amazon? Where had that come from?

Back then, the plight of the Amazon was just beginning to percolate over the international news wires. The Brazilian government was being criticized for policies that allowed for unchecked development. Read: cut down the ancient trees, plow up the ancient soil, and raise beef cattle. The newspapers reported dark, apocalyptic smoke clouds stretching for miles over primeval jungles subjected to slash-and-burn techniques. People who were trying to stop it were being assassinated. I had watched a few stories about this on various television programs, and it was something that really concerned me.

I kept writing.

I want to ski the Rockies. I want to ski until my legs won't work anymore.

I've always been an avid skier and I guess I'd always felt that you couldn't really call yourself a skier until you'd skied the Rockies.

Anything else? I thought. No. I sat for a while at my table, but no other goals came out of my pen.

Finished, I folded my paper, left it on the table, and went to bed.

The next morning I got up, made coffee, and read my list over a couple of times.

Yep, I thought. This is still what I want to do with my life.

I didn't precisely know why I'd chosen these things, nor did I have any idea how to begin the process they seemed to demand. Still, I couldn't help but notice how none of the items on my list had anything to do with the management training program at IBM Australia. A few hours later, I went to my office and turned in my two weeks' notice.

A day or two later, the head of IBM Australia called and said, you should not do this. When I told him my mind was made up, he took me out to lunch as a way of talking me out of resigning. He told me how nice it had been when he was a director, what kind of a future it meant for me. He regaled me with stories of his own experience in the upper ranks of management. How he'd learned to be a leader. A better man. A better person.

All in all, it was a very nice lunch, and one that ended surprisingly. Once the waiter had cleared our plates, the head sat back in his chair and appeared to get lost in his thoughts for a moment. I thought he was gathering himself for another pitch, but eventually he shrugged and his features relaxed in a way I hadn't yet seen.

"You know," he said. "You're not the first. The head of our

Southern Australia branch is sort of my counterpart. Last year, he suddenly up and quit. Said the same thing as you. That it wasn't for him."

I just stared at him. He smiled.

"Listen," he said. "You're young. You've got to do what you think you should do. We'd love to have you on the team, but if you want to follow your bliss, I say, Godspeed."

I think about this man every now and then and appreciate what he did that day. He didn't try to hold me back and he didn't push me out the door. He stood in the threshold between my present and my future and, like a proper gentleman, he said, stay sharp, now, lad. This is all up to you.

I got my affairs in order and bought a one-way ticket to the United States.

It was ski season when I hit the Rockies, and I hit them hard. Aspen. Vail. Jackson Hole. Steamboat. Alta. Snowbird. Over the course of seven weeks, I skied every slope I could find. I slept in the backseat of my rented car a few times. While it might not sound glamorous, I felt like a king. The rest of the time, I crashed on the floors and couches of friends vacationing in the region. I was twenty-eight years old and I had no agenda or preconceptions. I never knew where my next meal was coming from or what I'd be doing from one moment to the next. It was heaven. The time of my life. But even this must end.

Toward the end of my seven weeks, I found myself mulling over the piece of paper I'd written on at my kitchen table.

The rain forests, I thought.

So I booked a flight to Manhattan.

My friend John McConnell had recently graduated from Wharton with his MBA. A big bank had snatched him right up; it was Merrill Lynch or Goldman Sachs, I forget which one. He and his wife had just had a baby. They were living in an apartment on the East Side, a stone's throw away from the United Nations. His company had rented the place for them until they found proper housing. John knew I was looking for contacts in the environmental sphere, and he offered his pull-out couch for a few nights. Quite generous of him, really. Doubly so in light of the odds I faced. Having no contacts, acquaintances, nor even someone to vouch for me, I resorted to cold-calling NGOs.

Step one: pick up the phone. Step two: dial the number for the UN, Conservation International, Wildlife Conservation, the WWF, whoever was next on the list I had made. Step three: introduce myself, making sure to keep my pitch simple.

"Hi! My name is Rich Fuller. I ran a nonprofit called the Tree Project in Melbourne, Australia. I'm here in the U.S., and I'd like to work on saving the Amazon rain forest. Can I help you out in any way?"

If you've never thrown yourself into the ring of life without boxing gloves or satin trunks, I urge you try it sometime. Sometimes, the line I called clicked dead. Or people said, sorry, we're not interested, best of luck. But fortune smiled on me once in a while. Here and there, I found kind, patient souls who would take a few moments and hear me out.

Of the dozens of people I must have called, a few offered me contact information for organizations based in the Amazon. These weren't definite leads, of course. More like, here are some numbers you could try and the addresses of people I correspond with. Contact these folks and introduce yourself. Tell them what you just told me, and who knows? Perhaps they can help you.

I followed up on each of these leads, and made some new friends in the process, though no one offered me a job. But then I got a lucky break. One of my calls led to an appointment with Dr. Noel Brown, who at the time was the head of the New York office of the United Nations Environment Programme. Noel was very kind to me. We met a few times, and he referred me to a guy in Brazil whom I'll call Jorge da Silva. Jorge, I was told, worked for a group called the Santo Daime in Rio de Janeiro.

"Jorge came to Manhattan recently," Noel said. "I met him. He's an interesting man. Apparently, he and his group live on a plot of land in eastern Acre, a tiny province in the heart of the Amazon basin. Deforesters are raping the land. I didn't know what help we could offer him, but listen. If you want to meet someone down in Brazil who is actively fighting this war . . . "

Noel Brown passed me Jorge's contact information.

It sounds a little ridiculous saying this now, but armed with little more than that, I ordered a bunch of tapes from Berlitz and started learning Portuguese while riding the New York subway. Within weeks, I had booked my flight to Brazil.

So much for punching a clock.

CHAPTER TWO

Up the River to Find Its Source

I called on Jorge da Silva in Rio with nothing more than Noel Brown's business card as my entrée. Since I offered the card with gravitas, and since my Portuguese was too crude to express much to the contrary, Jorge and his fellows assumed that I worked for Noel at the UN, and I didn't do anything to dissuade them of this notion. It was easier than trying to explain that I had flown to Brazil on my own dime. That Noel Brown and I had only met a couple of times. That I'd recently worked for IBM selling telephone systems to enterprises both private and public. That I'd left all that to ski the Rockies for seven weeks. And so on.

"So listen," I said. "I hear you've got this rainforest land that's threatened with deforestation. What can we do to stop that from happening?"

Jorge put me up at the Santo Daime compound in Rio. It was an eye-opening experience. No one had briefed me on the fact that Santo Daime is a religion whose participants mix Christianity with the ancient American Indian ritual of ingesting psychedelics. Specifically, Jorge and his fellows ingested ayahuasca, a yucky brown tea made from boiling the *Banisteriopsis caapi* vine and the leaves of the *Psychotria viridis* shrub. *Daime* is a slang word for ayahuasca. It also means "give me" in Portuguese.

Therefore, Santo Daime literally means "saint give me."

According to Santo Daime tenets, drinking ayahuasca while focusing the mind through prayer can bequeath a worshiper with literally anything he or she desires. In more practical terms, drinking ayahuasca instigates hallucinogenic episodes complete with visual and auditory stimuli, wild emotional swings, a feeling of elevated consciousness, and, of course, *la purga*: fits of diarrhea and vomiting.

A few days after my arrival, Jorge escorted me to my first Santo Daime ceremony, where I took the ayahuasca sacrament. The rite took place in a big open room with a cross in the center and six zones organized hexagonally around it. One zone was designated for married men, another for married women, then single men and single women, and, finally, boys and girls. Each zone had simple wooden benches that ran four or five rows each.

I was given a small cup of ayahuasca to drink as I entered the room. The brew was bitter, not at all pleasant to swallow, but could be done quickly if I forced myself. After I had consumed the tea, curates escorted me to a specific spot in the tier and zone I belonged in. They gave me a hymnal that contained sweet little tunes and directions for dances to accompany them. The routines reminded me of kindergarten games. *Bah bah bah*—three steps to the right. *Bah bah bah*—three steps back to the left.

The idea, I think, was to loosen us up and give us something simple and fun to concentrate on. One at a time, each tier was led to the center of the room. Standing by the cross, resplendent in ceremonial robes, the high priest doled out shot glass after shot glass of ayahuasca. We downed our sacraments quickly and then returned to our spots, where we sang and danced until we were called to come forward again.

I felt privileged to take part in this ritual, and also somewhat prepared for it, given a few previous experiments I had made with LSD and psychedelic mushrooms. The effects of ayahuasca were more like my experience with mushrooms. At first, I thought nothing was happening. Then I noticed how tight my muscles had become, how my jaw had clamped shut. Looking up, I noticed that the colors of everyone's clothing, skin, and hair had begun to throb and breathe. At this point, I became aware of flashing lights and colors around the chamber. Thoughts became physical objects that I could reach out and touch with my fingers, or the tip of my nose. The effect of ayahuasca was strong and not at all unpleasant, except for the nausea. That was the main difference with ayahuasca. It doesn't matter how many trips you've taken on any other drug. In the end, everyone gets the runs.

Picture two or three hundred people swaying back and forth, chanting. The smell of their sweat stings the air like a smack in the gob. Some raise their hands with their heads thrown back and mouths agape, as though rapture is a medication best taken orally. Some close their eyes with their chins slammed down on their chests, humbling themselves before God. But their voices, as one, ascend in song.

Daime forca, daime amor, daime luz!

Give me strength, give me love, give me light!

During all of this, the priest and several novitiates wandered through the masses. They passed out tissues, comforted those who were weeping or scared, and led more than a few toward the toilets staged outside the hall. Congregants occasionally fled the room to collapse in the hot sun outside the building.

Again and again, we went up to the altar. Again and again, we downed shots of ayahuasca and returned to our spots. In no time at all, my mind slipped free of its moorings and started to bounce off the guardrails of infinity.

What am I doing here? Why is that wall bleeding luminescent colors? Who am I that claims this mind, and who am I really, since names are mere masks and identities are leaves that fall to the ground in the autumn of life? My essence belongs in heaven, I thought. In heaven, there are pancakes.

Hang in there. Don't lose yourself. Deep, slow breaths. Feel your feet on the ground. Feel your center within you.

Daime forca! Daime amor!

As árvores!

The trees!

Over the next two or three weeks, I attended the ayahuasca ceremony several times, after which I was invited to Santo Daime's parent church, deep in the Amazon jungle. Jorge warned me that it would not be an easy journey. We boarded an ailing VARIG jetliner that spent five hours bouncing us toward South America's interior in forty-minute jaunts. It was basically a jungle commuter flight. The only thing missing was a clutch of live chickens squawking from cages stashed under some peasant's seat.

At the final stop on the VARIG line, Jorge unfolded a map of Brazil and began stabbing locations with his finger. He showed me how we had flown northwest from Rio de Janeiro, over the states of Minas Gerais, Goiás, and Mato Grosso. Now we were in Rio Branco, the capital city of Acre, Brazil's westernmost state and a victim of rampant deforestation.

We switched to a dilapidated four-seat Cessna jet that jerked us into the air, belching smoke and soot from its engines. The Cessna took us to Boca do Acre, a dingy village bordered by the green-grey umbilical kink of the dingier Purus River. At this point, one could fairly say we had passed beyond civilization. However, we were still a long day's journey from the Santo Daime compound.

Boca do Acre was nothing like I had imagined a place so deep in the Amazon would look like. It consisted of just a few dirt roads and only one building with more than two stories—our hotel, in fact, which had three floors and ten rooms to let. The river flowed just outside the front door of whatever building you were in. Boats of various sizes punted up and down it. Jorge and I ate at the only restaurant, which served only one meal, that being whatever *senhora* was cooking that evening.

The next morning, while it was still dark, we met by the edge of the Purus and boarded Jorge's speedboat. He told me that someone had deeded the craft to the church; I presumed that this occurred while the grantor was high on ayahuasca. From my own experience with the psychedelic tea, I would expect that such flights of generosity were not uncharacteristic.

The boat didn't strike me as particularly seaworthy, but Jorge cast off, gunned the engine, and steered us into the river. Within a few minutes, the scenery changed. The settled shores gave way and I could feel the jungle closing in on either side, and above us, as well. The screech of monkeys off in the woods and the flap-flap-flutter of brightly colored parrots filled the air. Small moths flitted across the surface of the brown water in droves.

There's no one out here but us, I remember thinking. This is the jungle. I'm in the jungle.

When dawn finally slipped over the horizon, Jorge blasted the Beatles' "Here Comes the Sun" over the stereo and passed me a monstrous marijuana cigarette, which we smoked while he revved the engines. We were slamming up the Purus now, skipping the surface like a flat stone bouncing at thirty miles per hour.

A few hours later, we put in at a small hut that stood at the confluence of the Mapiá. Where the Purus had often ranged 200 meters wide, the Mapiá narrowed to little more than an *iguasu*, or stream, which the speedboat couldn't navigate. Jorge turned it in for a canoe equipped with a propeller on a long stick poking out from the back. Those poles are great because you can lift them out of the water quickly to clear obstructions or, if you like, slam them to one side and spin your craft 360 degrees in the blink of an eye. At this point, a couple of very experienced guides took over and handled the driving.

I sat on a hard wooden bench in the center of the craft with my bags in a pile at my feet. My chief fixation had become the safety of my Epson laptop, a dinosaur by today's standards, big as a briefcase and cinderblock-heavy. Rain forests are not hospitable places for delicate electronics. They're enormously hot, with humidity ranging close to 100 percent. It often rains four or five times a day. No matter how hard I tried to keep the computer dry, at one point it fell into the river. When we hauled it out quickly and cracked the case, I was relieved to find that water had barely permeated it. Still, the thing became riddled with insects. Quite often during the length of my trip, I would hit a key and hear the soft crunch of a bug's shell being crushed. And yet, for all this, that Epson never let me down. It was the draft horse of computers, and I still think of it fondly.

With its shallow hull drawing so little water, our canoe zipped straight up the Mapiá. At some point, we left the state of Acre behind and crossed into Amazonas, the largest state in Brazil, spreading all the way up to the Andes and including most of the Amazon basin. The jungle pressed down on us, crushing in from both sides with arms of vine, trunk, and canopy. Just when I thought we were due to be smothered, at the point where the Mapiá itself seemed to have lost itself in the bramble, we steered around a deadfall dam and into the proceeding bend where, miraculously, the forest gave way to open water.

"What is this place?" I said out loud. I couldn't believe my eyes.

Jorge tapped me on the shoulder and grinned. "Welcome to Céu do Mapiá," he said.

Heaven on the River.

Céu do Mapiá was a beautiful place, consisting of two clearings carved from the heart of the jungle on either side of the river with a crude bridge built to connect them. The village boasted a couple dozen simple houses, plus another Santo Daime church like the one I'd attended in Rio. As Jorge explained to me then, the colony had been founded a few years before by a mystic named Sebastião Mota de Melo, known more widely as Padrino Sebastião, who was the leader of Santo Daime. Padrino lived here, in the forest, with a large extended family that included Jorge by marriage. Shortly after we landed, Jorge introduced me to a young woman, a daughter of Padrino, that Jorge called his wife. This confused me since, back in Rio, Jorge had introduced me to another woman he called his wife.

"That was my wife back there," he said. "And this, you see, is my wife for here."

The colony, I was soon to discover, numbered barely 100 people, most of whom lived in thatched huts and kept chickens and pigs that wandered about at their leisure. The settlers had cleared a few plots to grow manioc, which I knew as cassava. They had one two-story dwelling, which belonged to Padrino Sebastião. The height and relative fortitude of the structure reflected his status in the outpost. They had also built a crude church where the entire population of the village—men, women, and children—gathered to take ayahuasca. Those who lived in the village took part in the sacrament; that was the unwritten rule.

I stayed in Céu do Mapiá for a month or so and studied the town's predicament. Noel Brown had been right when he hinted that loggers and farmers had recently stepped up their pace. Each year, their enterprises cut an area the approximate size of England, Scotland, and Wales combined from the heart of the Amazon jungle. There was no question that these operations threatened Céu do Mapiá. To my mind, however, they also spelled the extinction of something greater. Even back then, it seemed clear to me that, if left unchecked, the forces of "progress" could plow the entire Amazon under, destroying the largest and most biologically diverse rainforest on earth.

At the end of my second or third week there, I happened upon an idea.

"Who owns this forest?" I said.

Padrino Sebastião shrugged. "No one owns it," he said. "The forest is for everyone. The forest is ours. The forest is God's."

It was a lovely sentiment, and one I still uphold in its purest sense, but it wouldn't stop a bulldozer. We needed more earthly resources.

Back then, Brazilian environmental law was rudimentary at best. Under the Portuguese system, which Brazil still followed, the forest was considered Crown land—owned by the state until someone claimed it. That was good news because, as it turned out, there were also ways to claim the land and keep it safe from the bulldozers. These methods had been tested in other parts of *la floresta*. I felt certain they would work in Céu do Mapiá, as well.

"Here's the idea," I told Padrino Sebastião. "We write a plan to classify the village and all its surrounding land as a park—an extractive reserve, to use the official parlance. This petition will acknowledge the government's title to the land and resolve not to question it. But it also requests that you are written into the title as the forest's official guardian, that you are legally recognized as a cooperative group that maintains, nurtures, and protects the land in perpetuity, endowed by government-granted rights to ensure the strength of your mission. We'll include a clause in this language stipulating that no one else can do anything to the forest without your approval, including cutting it down."

Padrino Sebastião looked at Jorge. "A park?" he asked.

"Exactly," I said. "With you as custodians."

Using my Epson, I wrote the proposal to create an extractive reserve. It was basically two documents, each the approximate length of a business paper, and included draft regulations that would have to be passed, some procedures to follow, a sample charter for the guardian community to follow, and a breakdown of total costs, which were modest. The whole scheme, I wrote, would likely cost less than $100,000 to implement. I based all this work on precedents I had read about that were set by similar reserves in other parts of the Brazilian rainforest.

When I was finished, I showed my work to Padrino Sebastião, who read the petition and said he liked it. Moreover, he said that I should take the plan at once to the region's representative in the government—a woman named Sadia Hauache, who lived clear across Amazonas in Manaus, the state capital. Having no means to communicate from Céu do Mapiá, I left the jungle armed only with my petition, Sadia's name and address, and a brief note scratched by Padrino Sebastião. Not that I would need much more.

Back then, Manaus sprawled like a spent burro on the confluence of the Negro and Solimões Rivers. There were no roads in or out; visitors could only arrive there by boat or plane. Once I arrived and explained who I was, Sadia and her son, Abdul, accepted me with open arms, since Padrino Sebastião's word was better than gold. Sadia and Abdul read my petition and loved the idea.

"What are you doing next week?" Sadia asked.

I told her that my calendar looked free.

"Good," she said. "I have some meetings with the federal government in Brasilia. I'll get you a ticket. Come with me."

I spent most of the next few days searching for a printer. The one I eventually purchased cost me a fortune, but what could I do? Nowadays, you can find a Kinko's or Staples on practically every corner of every small town from here to Timbuktu. Heck, I bet there's a Kinko's or a Staples *in* Timbuktu, with an OfficeMax and an Office Depot thrown in for good measure. But back then? In Manaus? Finding a printer was like locating a pin in a mountain of needles. Finding paper that would feed into that dot matrix printer was even harder. In the end, I only found one store that carried it, and they only had one batch, which was supposed to last them until the next supply plane arrived, mid-year. When I bought their entire reserve, they looked both elated and deeply suspicious.

Our flight from Manaus took two or three hours. While we were in the air, Sadia explained how Brasilia was a designed capital. It didn't exist until 1958, when the Brazilian federal government wanted to move itself from Rio de Janeiro, closer to the heart of the country. Like a U.S. senator or congressman, Sadia kept an apartment in the capital for when she arrived to conduct her business. I found the way she described her job as intoxicating as the whirlwind pace we kept. When the plane landed, we jumped into a waiting car that barreled past gorgeous buildings built in the white-winged style perfected by Oscar Niemeyer, the architect who designed most of the city. I would have loved to stay and stare all day, but Sadia meant business.

At one of the main government buildings, Sadia walked me past the doors of five or six sub-secretaries, straight into the office of the minister of land. I suddenly realized I wasn't wearing a tie and felt like a fool. I can't remember the minister's name, but I recall how he rose at once, embraced Sadia, and blew kisses on either side of her face before turning to me and extending his hand. His Portuguese rattled out far too quickly for me to follow. Sadia's speech, I noticed, changed at once to match what I gathered was the patois of the capital. But I heard her introduce me as her friend Rich from the UN.

"He has a plan to save the rain forest," she said. "A good plan. Take a look."

Back in Manaus, I had printed out eight complete copies of my petition, one of which I now handed to the minister with the most officious air I could muster. He accepted the thick packet graciously, glanced at the title, and read the first few lines before nodding.

"Great," he said. "What are you doing for breakfast tomorrow?"

"We're free," I replied.

The next morning, Sadia, the minister, and I had breakfast with José Sarney de Araújo Costa, the president of Brazil. I still was not wearing a tie. The Brazilians did most of the talking, at any rate. I could barely make out a word they were saying; their Portuguese was far more advanced than mine, and they plied it at the breakneck speed mastered by seasoned political operatives.

At one point, however, President Sarney turned to me and said, quite clearly, "This is what we need." He held up a copy of my proposal. "More like this. Can you do that for us?"

"Yes," I said in Portuguese, nodding. "We can make these parks all over the country. How many would you like?"

As I later came to understand, the massive pressure Brazil was facing from the international community worked in my favor. Everyone wanted the government to halt deforestation, but the government had no concrete plans. By luck or fate or whatever you call it, I was one of the first people to bring them something actionable, legislation they could point to and say, "Look. You see? This is what we're doing, okay? This is our response."

The president told me to meet with his minister of the environment, a man named Fernando Mesquita, which is what I did the very next day. Senhor Mesquita read my proposal and smiled.

"Yes, yes," he said. "This is fine. You wrote this quite well for an outsider. It will have to go through channels, of course, but I'll start the process at once. Let's see how quickly we can make it happen."

He was as good as his word. The proposal was signed, sealed, and stamped into law within the year. The village of Céu do Mapiá and its surrounding regions became a park under federal law, with a team of chief custodians at Céu do Mapiá. Mesquita sent me a nice letter thanking me for everything I had done. By that point, I was already back in New York and reporting what I had done to Noel Brown.

Feeling a little awkward, I said, "Hey. Remember your business card? Well, I passed that around a bit and guess what? I ended up establishing a national park in the Amazon rain forest. Now the Brazilian government wants me to do more. I've already done some research and I think I can prove six more viable reserves at the locations I've marked on this map. Do you want to help?"

Noel got very excited. "This is fantastic!" he said. "I have to go down to Brasilia in two weeks. Come with me!"

So I did.

At a conference in Brazil, Nocl took me around and introduced me to lots of people I otherwise wouldn't have met. Later during that trip, I got Noel a meeting with Senhor Mesquita. Noel was over the moon about that. Though he had tried, he'd never been able to score himself a one-on-one with Brazil's minister of the environment.

When the conference in Brasilia was finished, Noel flew back to New York while I made a detour back to Manaus to meet with Sadia Hauache and update her on my progress. On her dining room table, I spread out my map of the Amazon and ticked off the sites I was planning to visit.

"These spots," I said, "are the ones that I feel best qualify for extractive reserves."

Sadia agreed. All the places I'd marked fell within her constituency. Her husband knew the area very well. Over the next few days, she reviewed my plans and gave me frank advice on which sites would welcome me, and which would not.

"You cannot assume that each effort will go as smoothly as Céu do Mapiá did," she said. "You arrived at that village as a guest of the Santo Daime. In more remote regions of the rain forest, you will enjoy no such overture. The people who live in these places can be suspicious of outsiders. You must take great care with your words and deeds."

She promised to hire a riverboat that would take me into the jungle when I returned. I was overwhelmed by her generosity, but Sadia shrugged.

"It will make traveling easier for you," she said. "And safer. Return to New York, but come back to us soon. Your work here is only beginning."

I didn't want to go back to New York, but the small fortune I'd made at IBM looked dismally smaller each time I checked my bank account. At that time, I was still doing practically all of this work on my own dime. I would have welcomed an infusion of capital. At one point, Noel Brown had offered me a consulting contract, and I wanted to pin him down on the details before I found myself begging for change in the Times Square subway station.

But when I met with Noel again, he apologized.

"Funding is scarce at the moment," he said.

In lieu of a paid position, he offered me a desk in his office at the UN. I accepted, although the arrangement felt strange. Had I not proven my legitimacy? How many UN employees, I wondered, had flown to a foreign country on a whim and their own dime, and wound up having breakfast with the country's president who then empowered them to save large tracts of the Amazon rain forest?

Despite my success, I had no income, was crashing on couches again, and was losing friends in the process.

But things got better, little by little. I won some grant money from an environmental group up in Boston, which meant I wouldn't starve. Then Noel's assistant lost her roommate suddenly and offered me a space in her Lower East Side apartment for reasonable rent.

Okay, I thought. Clearly this situation isn't everything I want, but it certainly feels like I'm moving in the right direction.

After a couple of months' hiatus, I returned to South America to finish the job I'd started.

Sadia made good on her word and hired a riverboat for me. It was one of those two-story jobs with a hull as thick as a bank vault door, and with damn good reason. The Amazon is a treacherous river, filled with all kinds of destructive debris. An extra-thick hull can absorb almost any impact. Low-hanging curtains of bramble can't hold it. It can push through downed trees and capsized canoes. Such boats essentially stop for nothing. Like lazy hippopotamuses, they putt-putt-putt against currents and weather to ferry you slowly upriver.

During this leg of my journey, I often felt like I'd landed on the set of *Apocalypse Now*. Granted, no one was shooting at me (a fact which I deeply appreciated), but I passed many hours lounging in a frayed hammock strung between deck pilasters. The boatmen chattered constantly in a dialect I gave up trying to parse. We drank beer and whisky and the strongest coffee I'd ever tasted—thick, black sludge, which we made by boiling hand-crushed grounds in an old tube sock for half an hour.

We smoked pot occasionally because there was basically nothing else to do. Besides, it made perfect sense to stay numb since it rained twice an hour. The air felt heavy as iron, the heat and humidity sank into our bones, and bugs scratched nests in our nostrils and armpits. Meanwhile, the forest primeval rolled past as thick as castle walls on either shore. Now and then some vine or branch would reach out, grasping with leafy green claws.

The jungle, I thought, is trying to yank me out of this boat. It wants me, though God only knows why.

We had already ventured deeper into the Amazon than I had ever been in order to reach Céu do Mapiá. Now we were going deeper still, and our journey had barely begun.

I passed nearly two months like this, putt-putting up and down the Purus, two trips in all. We stopped at the towns I'd planned to visit and several I hadn't, since they weren't on our maps. The people who lived there were welcoming in a standoffish way, happy to see strangers but also deeply suspicious of my motive for arriving in such remote places in a big boat. At every location, I presented my UN card, but their moods only brightened when I mentioned I worked with Sadia Hauache. Even in the most remote regions of the Amazon, her name was a gold standard.

Once my credentials were established, I asked to meet with the local mayor and explained my plan to propose their settlement as an extractive reserve. In every case, I asked how far the townspeople had walked into the jungle from their homes. Though somewhat crude, this was the means Sadia and I had decided should establish a perimeter for custodianship.

If any of this makes you think that I was winging it, you're absolutely right. To be clear, I'd started making things up as I went along the moment I left IBM. But I was getting results and having the time of my life, so I saw no reason to stop.

By the time my second trip was through, I had written proposals for four more extractive reserves, all of which I think eventually passed into law. In total, my efforts probably ended up preserving an area of Brazilian rain forest equal to half the size of Massachusetts. I say "I think" and "probably" because, at this point, something awful happened that kept me from knowing the true results of my work.

Being young, enthusiastic, and an outsider, I didn't understand the political consequences of what I was doing. In my mind, I was acting for the greater good: saving wide swaths of the Amazon rain forest, one of the most cherished natural resources on earth. In the minds of certain others, however, it was not so. To them, I was this UN guy, an outlander and a white kid who had arrived in Brazil out of nowhere and gone over everyone's heads. By speaking directly to the president and his minister of the environment, I had not respected local channels. Hell, I hadn't even acknowledged that such channels existed.

Let me give you a better idea of who these people were. A lot of them carried guns and very large knives and ran certain enterprises out in the jungle that authorities in more civilized places would surely have liked to curtail. Whatever law exists in the jungle is the law that these people approve. By their way of thinking, I wasn't preserving the land so much as stealing it from them bit by bit and placing it into the hands of the federal government they despised.

The person who explained all this to me was named Tomaso. He worked at Mesquita's office in a town called Rio Branco, on the edge of the Amazon. I often used Rio Branco as a base of operations, and during one of my stops there, Tomaso calmly explained that there was a price on my head.

"My friend," he said, "you should leave here at once."

Tomaso said he had overheard several men talking openly about killing me.

"That's ridiculous," I said.

"You should leave," he urged again.

Like I said, I was young, enthusiastic, and an outsider. I was about to learn a hard lesson.

The next evening, I went to a restaurant I'd eaten in several times, a little shack that smelled of herbs and *feijoada*, the classic Brazilian pork and bean dish. The place had a corrugated tin roof and tin walls that could be moved like shutters—very convenient for tropical climes. The tables were slabs of wood, thick as pylons, and the floor sported patterns of bright mosaic tile.

I remember it rained pretty hard that day. Well, it rained pretty hard almost every day, but this time it was still coming down when my group and I pulled up in our cars. Our party numbered six or seven in all: myself, Tomaso, the captain of my riverboat, and a few people who worked in Tomaso's office, all good people, all friends. We bolted from our cars and scampered into the restaurant as fast as we could, though we needn't have bothered. The moment we got under the roof, the rain shut off like a faucet. This was typical for Acre. If you didn't like the weather, all you had to do was wait a few moments for it to change.

We sat at one of the big tables in the back, near the kitchen door, and ordered a round of drinks plus some deep-fried vegetable appetizers. I remember that I had just cracked open an ice-cold Antarctica beer when the first course came out and we started to eat. Along with the *feijoada*, we had *pirarucu*, a prehistoric beast of a fish that, thankfully, tastes better than it looks. We were right in the middle of chatting and chewing when the restaurateur, who was also the chef, burst out of the kitchen looking very harried, and started rattling in Portuguese, too loud and fast for me to follow.

"What is it?" I asked (or started to, anyway).

Tomaso jumped up, grabbed my arm, and dragged me toward the kitchen. I launched into a series of questions, protesting, but Tomaso wouldn't stop. He pulled me through the kitchen door, past the hot stove with its clutch of pots bubbling on the range, out a back door, across the parking lot, and into one of our cars.

The rest of my party was right behind us. Over my protestations, someone keyed the ignition of the vehicle I was in and we all drove away before the doors had even closed. By this point, I felt stupid and scared.

"What is it?" I asked again in Portuguese. "What's happening?"

We sped to a house owned by one of Tomaso's friends and went inside. Beers were plucked from a refrigerator and passed around, and again I asked what had happened. This is what my friends told me:

The chef had been working at his stove when he looked up and saw a car pull into the parking lot. Some men got out and they all had guns, which wasn't really so odd. Back then, I used to see pistols all the time in that part of Brazil. I was never comfortable with it, but then no one ever asked my opinion.

At any rate, the chef recognized these men as part of a group not known for its diplomacy. He went out to ask them what was going on. The men with the guns said they wanted to see the Australian guy from ONU, meaning L'Organisation Internationale des Nations Unies. Meaning the UN. Meaning me.

The chef was a good man. He told them he didn't know where I was, so the men with guns moved off down the street while the chef slipped back inside, went straight to our table, and explained what he'd seen. Which is when Tomaso had hustled me out.

I remember hearing all this and thinking that my friends had lost their minds.

"I told you," Tomaso said. "You must leave."

This time, I didn't argue. I had them drive me back to my tiny hotel, where I went up to my room on the third floor, locked the door, and lay down to get some sleep.

I spent the next day in my room, on my Epson. No big events occurred. I got good pages done on the latest proposal I was writing. But the next morning, somebody knocked on my door at about eight o'clock. I got up from the little table I was working at and went to look through the peephole.

It was the hotel manager, a man I knew fairly well since this was my sixth or seventh stay at his place. He looked upset, so I let him in.

"You have to leave now," he said. "My friend just called from the airport. A man is there. He's drunk and waving a pistol around and screaming that he's coming to kill you!"

Again, I thought this was crazy.

"I haven't done anything wrong," I said. "Can't we call the police or something? Hey!"

The manager had barged past me into my room and started to dart around, gathering clothes and books, which he threw on the bed.

"I have a car," he said. "This man? My friend said he was stark raving mad, demanding to see the tall Australian from ONU."

I thought of what had happened at the restaurant the day before last, and that was all it took. Grabbing my suitcases, I stuffed them with clothes. Then I figured to hell with that, clothes are replaceable, flesh and bone is not. So I snapped up my Epson and followed the manager out of the room, down the fire stairs, to an alley behind the hotel. He waved me toward his ancient Mazda sedan, his personal car that was also his cab. Like many merchants in remote locales, the manager had more than one job. He moonlighted as a taxi driver.

"Get in the back!" he said. "Get down and stay out of sight!"

This was easier said than done. In the first place, I stand about six feet tall. More importantly, however, the floor of that sedan had rusted out long ago. I ended up draping myself over the torn, sprung upholstery, clutching my Epson to my chest in hopes that it would stop bullets, and staring down at the dirt track zipping past, a little more than a foot from my nose.

The manager avoided the main streets and blasted down back roads, angling toward the airport. We were pretty far out in the sticks, so the local facility was little more than a single landing strip on the outskirts of town. The place had a couple of Quonset huts set up to handle private aircraft. Each time I passed through there, I saw the same gaggle of ancient mechanics lounging on crates full of broken machinery, smoking and drinking beer. Not a rustic airport so much as a dilapidated one. I was never so pleased to arrive there.

The hotel manager screeched to a stop behind a primitive earthen berm. This was part of a new construction project. Bulldozers had pushed earth and debris into ridges like hedgerows, perhaps preparing the way for a new terminal. The manager told me to wait in the car and stay out of sight while he went to scout the place out. Luck was on my side again. Behind the berm, I was, for all intents and purposes, invisible.

A few minutes later, the manager returned and reported that, unfortunately, the local flight had just departed. It wouldn't return until the next day.

"But not to worry," he said. "My friend—the one who called me to warn you? He owns a Cessna and owes me a favor. He says he will fly you to the next town, which is not far from here."

"When?" I asked.

"Right now," he said, helping me out of the Mazda.

The ride in that Cessna was everything you can probably imagine it to be. The plane was old, held together with spit, twine, and electrical tape. At various points, I felt certain that the pilot's constant chatter was the only thing keeping wind under his aircraft's crooked wings. Whenever the Cessna dipped suddenly or its engine hacked smoke, which was often, I found myself mentally tallying pros and cons. Was it better to be murdered by drunken Brazilian assassins or to die in the fiery wreckage of a plane crash in the jungle? I never drew a conclusion on that one.

Fortune smiled, and we landed safely. At the next airport, I kept my head down while waiting for the next VARIG flight; this one went to Cuiabá, then Belo Horizonte, and finally Rio, where I stopped to gather my wits and find a phone.

I first called Sadia Hauache, but her secretary told me flat out that Sadia could not speak with me. This stunned and confused me, doubly so when Mesquita's office gave me the same response. Why was I being ostracized? It took me a while to figure it out.

As I've mentioned, I was young at the time, with a lightweight's grasp of certain political realities. As a tall, white outlander, I had always been conspicuous in Brazil, a predicament that grew starker the further I penetrated the jungle. In many places I visited, the people had rarely met anyone from Australia. They regarded me with a mixture of wonder and deep suspicion. I'd worked very hard to remove these hurdles. Using my broken Portuguese, I'd explained my mission to anyone who would listen. But there is a point where even the most motivated outsider, even one working informally for Brazil's federal government, could go no further.

The powerful people who ran these rainforest outposts benefited greatly from the status quo. They benefited from deforestation, and they had realized what I was up to. They had contacts and friends who owed them favors, and they'd called some in. I was done.

I was able to confirm this when I called up a confidante in Mesquita's office, a secretary with whom I would develop a lovely rapport. She seemed hesitant to speak at first. Without saying so directly, she let me know that I should not call again. Though I left a message for the minister to call me, I was not surprised when he never replied.

If all of this sounds paranoid, consider the way things were run in the rainforests of Brazil during the interval I was there. In every settlement I visited, the local land records office was never located in the village proper. Why? Because disputes over property would arise from time to time. When they did, it was common practice for one party or another to burn down the land office, destroying all the titles on file.

This was paper documentation, of course. During the early 1980s, Brazil was essentially a developing nation. Rain forest municipalities kept their records on paper. Computers were nonexistent. I still remember the look on people's faces each time I turned on my Epson and started to type. You'd have thought I had just grown a second head.

Abruptly, my adventures in South America had ended, and it was time to go home. There was only one problem. At that point, I had no idea where home was.

I could always go back to Australia, of course. I certainly had great stories to tell. But making a story wasn't one of the items on my to-do list. I had skied the Rockies until my legs wouldn't work anymore. Check. I'd gone to Brazil, learned what was going on with the Amazon rain forest, and managed (I think) to save quite a chunk of it. Check. If I didn't like how that adventure had concluded, at least I felt justified in scratching the item off my list.

So what did that leave?

I want to make a difference on a global level.

Hmm, I thought. Easier said than done. Let's table that one for the moment. What else?

I want to live in a foreign country, learn a different language, and experience a different culture.

Well, I thought, technically, the United States is a foreign country and a different culture. You can also argue that it uses a different language, since Yanks favor such a wacky version of English. Blokes are referred to as guys or dudes, and no one says "wacko, mate!" when they're stoked.

I was basically out of money. I had no job, no income, no visa status. My situation looked grim, indeed.

But not as grim as it looks in Brazil, I thought. At least in New York, I likely won't be shot.

Though it hurt me to leave at that time, in that way, I booked my flight back to the Big Apple.

CHAPTER THREE

How to Plant a Great Forest

So I ended up back in New York with my tail between my legs. It didn't take me long to find that my contacts at the UN didn't want to have anything more to do with me. Had they heard about the problems I'd encountered? Try as I did, I couldn't wrangle a meeting with Noel Brown. One of his assistants advised me to pick up my things, which they had packed in a cardboard box and left at the office's front desk.

I went to the office and camped out by Noel's door for a few hours, but my former mentor, once a fan, never appeared. I felt betrayed. The system had loved me while thing were going well, but dumped me with shocking speed the moment things soured through no fault of my own.

Again, I reviewed my options. I preferred to stay in the U.S., since returning to Australia would feel like failure. I had practically nothing to show for my efforts. True, the minister had all my proposals, but I was the worker bee; I wasn't destined for recognition when the final product became a success.

To complicate matters, I'd met a woman and fallen in love. We moved in together at a place on Twenty-Third and Seventh, and began to talk sincerely about starting a family. I knew I needed to get my act together, and fast.

I've always been inclined toward entrepreneurship. Why

not start a business? I thought.

With little idea what specifics I would engage in, I set up a corporation that I named Great Forest, after the Brazilian term for the Amazon, *La Grande Floresta*. Then I pulled out my dog-eared black book and began cold-calling every contact I'd made since arriving in New York. A few weeks later, I had my first lead.

Mark Marion had met with me several times during my trips back and forth from Brazil. When I first got to know him, he'd started a nonprofit rain forest action program, which sadly didn't last long. But Mark was indefatigable. By the time I caught up with him again, he'd founded a new company called Eco-Matrix, headquartered in the Lipstick Building at Fifty-Third and Third. It was a start-up in the truest sense of the word, working out of one of those serviced units where you pay a thousand bucks a month to have an office, a shared conference room, and equal access to the coffee fountain. Bernie Madoff ran his firm on the floor directly below Mark's. New York City: this was the place to be.

Mark had built his new company to produce a bio-degradable diaper.

"You have any idea how many disposable diapers are used in the world each day?" he asked when I got him on the phone. "You have any idea what these diapers are made of? Most take a hundred years to fully decompose. Course, nobody really knows that for sure, since nobody's sat around a landfill for a century and studied the damn things, right?"

Mark had great ideas like that—simple and effective. I asked him how I could help his business grow and felt him shrug on the other end of the line.

"No clue," he said. "Right now I'm up to my eyeballs in diapers. Pitch me something."

"We could add to your product line," I said. "How about selling recycled paper as well?"

Remember, this was 1989. Though it may be hard to imagine

today, recycled paper was sort of a novel idea back then. It wasn't until 1993 that more paper was recycled than buried in landfills. Even so, I read a recent estimate that ninety-five percent of all business information is still stored on paper in some fashion or another. Back then, Manhattan offices ran on paper like cars run on gasoline. And they still do, of course.

Mark loved my idea. We set up a deal where he paid Great Forest $100 a day to set up a division selling recycled paper. I worked for Mark's firm for about seven or eight months before the company folded. It turns out that a biodegradable diaper also leaks pretty badly. Plus, no one really thinks biodegradable plastics are such a good idea. Once again, I found myself out of work.

Okay, I thought. This is a setback, that's all. Something will turn up.

And it did—right down the hall, in fact. Another business on Mark's floor produced a line of shoes featuring hand-stitched beads in all sorts of garish designs. One of their best-selling products was a pair of moccasins with the American flag embroidered on the front in beads. These moccasins sold particularly well to NASCAR fans and the heartland crowd.

While working for Mark, I used to meet employees from this company at the coffee fountain. We chatted and got to know one another. When Mark announced that his business was folding, I must have looked pretty forlorn because one day, a friend who worked at the shoe company told me that they'd just lost their bookkeeper.

"If you're looking for work, I could put in a good word," he said. "Can you do bookkeeping?"

I told him of course I could. He brought me right into his office and introduced me to his boss, who asked me to stop by again tomorrow for a more detailed interview. After work that day, I ran to a bookstore, bought some manuals, and stayed up all night giving myself a crash course in business accounting.

The interview went well and I got the job, which paid ten bucks an hour. It wasn't much, but it paid some bills. I also kept my recycled paper business, which turned out to be important.

While working for Mark, I'd developed a list of clients and found a West Coast supplier for the recycled paper they needed. When Mark's company folded, I assumed ownership over this line of business. By doing so, I basically became the only guy selling recycled paper in New York City during the early 1990s. And people were buying it. Corporate environmentalism was still in its zygote phase, but developing somewhat. Big firms were leading the charge, mostly because they could afford to, and of course saying that they recycled made for great PR. New York Life became one of my customers, as well as a bunch of top Manhattan law firms, like LeBoeuf Lamb and Davis Polk.

If this sounds glamorous and profitable, let me correct you. I was miserable and broke. Yes, I was selling tons of product, but the margins on selling recycled paper were tiny. I only earned a dollar or less for every box of paper I sold, which is practically nothing. It's even less when you consider that I had to carry the cost of collections and all the logistical nightmares that come from functioning as a middle man. Half the time, I was frantic over whether or not a shipment would arrive in time. If it didn't (and this occurred frequently), I was the one who got hung out to dry, or who had to schlep several boxes of temporary supply across town so my clients could stand around chewing me out. Not that I blame them. They were well within their rights to do so. Imagine trying to run a law firm without paper. It can't be done.

I wasn't losing money, but I certainly wasn't making much. These days, people sometimes ask me what it's like to start your own business. I tell them: Don't even think about it. Go work for somebody else because, believe me, start-ups hurt. There's so much rejection, so little money, and so much stress that even the most driven personalities are likely to crack. A few years back,

I read a *Wall Street Journal* article that estimated how three out of every four start-ups fail. Indeed. Beware of starting your own business, I say. Almost certainly, you'll rue the day you thought you could work for yourself.

My job with the shoe company offered no solace, since it turned out that they were going out of business, too. To make matters worse, the FBI got involved. One day, a special agent contacted me and asked to see how I kept the company's chart of accounts. I met with him and his colleagues and showed them the methods I'd been instructed to use. The agents scanned my columns and frowned.

"What's this?" one of them asked. "A $30,000 tuition payment for a private school from the corporate checking account?"

"Yes," I said. "For the CEO's son. See the note in the margin? Those are my initials."

"You made this entry? That's not a legitimate business write-off."

"True," I said. "But it's not my job to counsel the CEO on which write-offs are legitimate and which are not. He told me to make these payments out of the company account, so I did."

The agents flipped through some pages.

"You've made more notes on these entries here," said one of them. "And here. And here."

I nodded. The CEO never checked the company books. I'd therefore made it a habit to note any transaction I considered suspect. I've said before that people who work in bureaucracies often make a career of covering their behinds. But that doesn't mean you shouldn't do it when you see something illegal taking place.

To make a long story short, the shoe business imploded in spectacular style. In the face of mismanagement, it was forced to liquidate its assets. The firm ended up dumping something like a million dollars' worth of merchandise on discount chains and dollar stores all over the country. The irony of this still slays me. Places like Neiman Marcus used to sell those beaded moccasins for over a hundred bucks a pair. Now you can pick them up on eBay for about five dollars and free shipping.

At some point during all this, my wife and I took a different apartment on Thirtieth and Third. I remember there were lots of crack dealers loitering in front of the building. Nice guys, though. They helped us unload our furniture when we moved in.

When my son, Max, was born in 1994, I realized once again that I had to diversify. In all honesty, I wasn't thinking about saving the world or making a difference in the grand scheme of things at that point. I was trying to feed my family, pure and simple. Still, it turned out to be a watershed moment for me. I could see all too clearly how, if I kept marching to the same tune and kept working in the same line of business, that my company would not make it. So I started brainstorming in earnest, and soon came up with what I thought might be a viable plan.

It started with a letter I wrote to my clients. In it, I argued that a truly great business makes a profit, obviously, but it also sees itself as integral to its community and the world. This is why truly great businesses insist on using recycled paper, I wrote. But why stop there? For a reasonable fee, Great Forest could become their environmental consultant. We could set up internal recycling programs, evaluate their firm's energy, water, and carbon efficiency, and recommend changes for maximization.

I'd never done this type of work before, but then again, neither had anyone else. To the best of my knowledge, the term "sustainability" hadn't even been coined yet. The same could be said for the notion of asking corporations to serve as stewards of the environment. In other words, this letter I sent was little more than a shot in the dark. But, once again, luck was on my side.

The venerable law firm of Sullivan & Cromwell became the first to accept my offer. I got a call from their office manager, who said he would love to have an in-house recycling program.

A law firm of Sullivan's size and quantity generates an enormous amount of paper waste.

"Currently," said the office manager, "we discard this waste as common trash. Which doesn't really make sense, if you think about it. What's the point of buying recycled paper if we don't continue to recycle it?"

I told him that I agreed completely. When he asked how much I charged and when I could start, I negotiated a $3,000 fee and told him I would start the next day. The amount of that fee was enormous for me at that time. Excited, I hung up the phone. I thought I'd just entered a new sector of business. But, as time would tell, I'd done much more than that.

Almost immediately, we began to pick up more and more contracts for environmental consulting. I'd be lying if I said we weren't lucky. Great Forest was a small, struggling company, yet we were already well-established when corporate environmentalism took off. We grew, and we grew fast. Within a year, I was able to rent an office on the fifty-first floor of the Empire State Building. I hired a few employees and had my company sponsor my green card.

One day, while reviewing our books, I compared the money we made from consulting to the money we made selling recycled paper. There was no contest. Margins for selling recycled paper were thin, doubly so in light of all the logistical nightmares. When one of my employees expressed interest in that line of work, I sold him my paper business for one dollar. I still consider this one of the most profitable trades I've ever executed. That dollar bought me time and peace of mind to focus on my consulting work. My bright now-ex-employee ended up reimagining the business as a printing/publishing outfit. No longer dependent on volume, he began to grow operations in several exciting directions. He's still in business today, and he's done very well for himself.

One referral led to another. Throughout the early '90s,

my staff and I found ourselves bouncing from one Manhattan building to another, setting up recycling programs. Though industrious, we kept a low profile, and with good reason. Back then, commercial garbage hauling in Manhattan was a $1.5-billion-a-year industry overseen by a mobster named James Failla, aka "Jimmy Brown." The garbage firms had to pay him fees to stay in business. In return, he ensured that there was no competition.

Jimmy Brown kept associates with colorful names like Joe the Cat, Joey Cigars, and Tommy Sparrow, men who were as hardcore as they probably sound. In 1989, Sparrow testified against the mob before a grand jury. So Jimmy Brown sent Sammy "The Bull" Gravano to murder him in a Brooklyn factory. Good times.

Restaurants, universities, hospitals, office towers—everyone did business indirectly with the mob. What choice did they have? Someone needed to take away your trash, and you could not get service from anyone except the guy allocated to your building. In 1992, a Brooklyn recycling company made a foray into the trash-hauling business. Mob muscle assaulted the firm's employees and firebombed one of its trucks. A year later, Browning-Ferris Industries, a nationwide hauling company, started taking New York clients. Until, that is, an officer of BFI found the severed head of a dog on his doorstep one morning. The dog had a note in its mouth that read "Welcome to New York."

But the price for acquiescing to the mob's demands was nearly as bad as the price for dissent. Back in 1994, a typical office tower in Manhattan paid about $0.60 per square foot every year to have its trash removed—almost double the rates found in commensurate cities like Chicago, Los Angeles, and Boston. At least one company I worked with, having been less than cooperative with their garbage company, found itself paying up to $1.20 per square foot. To give you an idea of scale, imagine one of those nice fifty-story towers on Fifth Avenue in midtown Manhattan. They typically run about a million square feet. Back in 1994, that building would have paid between $1 million and $1.2 million annually to have its waste removed.

Without question, haulers were gouging their customers, and much of that money went straight up the ladder to Jimmy Brown. John Gotti, the so-called "Teflon Don," was boss of the Gambinos in that era. Famously, federal surveillance once recorded him saying, "Jimmy Brown? He took the garbage industry and turned it into a candy store."

So that's the environment I was operating in. Was I nervous about it? Absolutely. At one point during the early '90s, a client hired me to consult for his business on Fifth Avenue. He wanted me to help reduce his monthly trash hauling costs, which is what recycling programs are supposed to do. I went to his building and conducted my study. By the end of it, I ascertained that his firm should probably pay, at most, $6,000 a month for trash hauling services.

"How much are you paying now?" I asked.

"The haulers get $25,000 a month," he said.

"Okay," I said. "Let's meet with the hauler."

Standard procedure would have me sit with the hauler and client to negotiate fair pricing. But this hauler's name was Bobby Franco, and he was an associate of Jimmy Brown's. On the day of our meeting, he called up to the conference room and said:

"I'm not setting foot in that office. You want to talk, you come downstairs. Meet me in the basement."

Yeah, right.

I said, "Bobby, there's no way we're meeting you in the basement."

"Fine," he said. "Then we're not doing business."

The client was scared stiff, and I can't say I blamed him. But something had to be done. If nothing else, my reputation and that of my business was on the line. So I went down to the lobby and talked to one of the doormen I'd met, a big weightlifter type named Louie. Louie's body looked like a bunch of cinder-blocks stacked inside his blazer. He was easily 300 pounds, and little if any of it was fat.

"Louie," I said. "Come with me, okay? I've got to go down and negotiate."

Louie shrugged and squeezed himself out from behind the concierge desk. Together, we went down to the basement parking garage, where Bobby Franco's black Cadillac stood idling at the loading dock. As we approached it, one of the tinted rear windows powered down. Bobby looked at me like I was a shit smear on the bottom of his shoe. Then he grinned.

"Get in the car," he said.

"Bobby, I'm not getting in your car."

"You get in the fucking car."

I turned to Louie and said, "You get in the car."

Louie looked pained. "I don't really want to," he said.

"Please, Louie," I said. "Just get in the car."

So Louie opened the door and put about a fifth of one of his butt cheeks on the back seat, which was basically all that would fit. Then I went round to the opposite side and tapped on the opposite window until that one, too, powered down.

"Okay," I said. "We're in the car. Now what is it?"

Bobby looked pissed. "What number you want?"

"You got my report," I said. "Six thousand a month."

He laughed. "I'm charging twenty-five right now. How can you tell me six?"

I said, "Six is the maximum, Bobby."

He shook his head. "Six from twenty-five? Cheese!"

I said, "Forget twenty-five. It should be six."

He shook his head again. A long silence passed.

"Fine," he said. "I'll do it for ten."

I accepted that. Ten. It was outrageous, of course, but it still generated big savings. My client went from paying $300,000 a year to $120,000 a year, and for this, I charged him $5,000, so everyone was happy. Well, everyone except Bobby Franco, but it turns out that his days were numbered anyway.

Not all of my interactions with the mafia went so swimmingly. On another occasion, I was leaving a client's office when a large man with slicked back hair and a leather jacket bumped into me a little too forcefully. When I said excuse me, he grinned and thumped a hand on my back.

"Heya, Richie. I hear you're going away for vacation. Next week, that right? With your wife and kid?"

I stared at him.

"Who are you?" I asked.

Still grinning, he opened his coat and showed me what looked like a gun holster.

"Better take care of yourself, you hear? Don't want nothin' to happen to your family, right?"

I might have quit right there, had it not been for Johnny Dellito.

Johnny was a hauler whom I'd been introduced to by a mutual friend. We'd golfed together and, once, he'd taken me for a ride in his Porsche 911, of which he was very proud.

Given his name, manner, and occupation, I suspected that Johnny had ties to the mob, but he never volunteered this information and of course I never asked. As luck would have it, I bumped into him a few days after I'd been threatened. We passed a few minutes in small talk, after which, apropos of nothing, Johnny said, "Listen, Richie. Don't worry about a thing, okay? I'm looking out for you."

Again, I found myself staring.

"What are you talking about, John?"

He shrugged. "I told them," he said. "You're small-time, not threatening anybody, right? So I smoothed it over, no big deal. Okay?"

"Okay," I said. Because, at the moment, this seemed like the only word in the English language.

He shook my hand and walked away, saying we'd have to go golfing sometime soon. I stood there watching him leave and wondered what the hell I'd gotten myself into. More specifically, what was real and what wasn't? Had I really just bumped into Johnny accidentally, or had our little meeting been planned? And why had he befriended me in the first place? Because his long game beat mine on almost any fairway? Or had he been scoping me out the whole time, trying to ascertain what kind of risk I posed to people I'd never met and the organization they represented?

To this day, I'm not really sure what kind of deal got cut behind my back. Truth be told, I don't think I want to know. At any rate, I kept at my business, kept a low profile, and no one threatened me ever again. And luck was about to cross my path again.

Within a few months, Robert Morgenthau, the district attorney for Manhattan, opened a landmark case that brought down the mob. His 114-point indictment accused seventeen individuals, twenty-three carting companies, and four trade associations of scamming the city for close to fifty years. Later, I was surprised to learn that I'd helped the investigation, however inadvertently.

One day, I read a copy of the indictment, and there was my name in black and white. Morgenthau's agents had tapped my phones without telling me, both at my home and at the offices of Great Forest. I read transcript after transcript of me offering my services to set up recycling programs and help reduce their inflated trash prices.

This is what trash hauling should cost, I would say. This rate is reasonable based on maximum rates. And this rate is not.

Morgenthau and his investigators had been watching me. They compared the prices I negotiated with those offered by certain Mafiosi who'd talked to undercover agents wearing wires, or whose phones had also been tapped. In some of those recordings, Mafia soldiers could be heard admitting that my prices were hurting their bottom line. They also discussed what should be done with me for daring to transgress against them.

The DA kept me out of the loop to protect me, I suppose, but also, I have to assume, to help him build his case. As everyone learned later on, he'd been amassing evidence against the mob for years. To this day, I'm thankful he never contacted me. If I'd known what he was doing back then, I'd have been scared out of my wits.

The syndicate finally came smashing down in 1997. Morgenthau's indictments sent seventeen key players to jail and broke the Mafia's stranglehold on waste hauling. Almost immediately, a massive upheaval shook the industry. City supervisors rooted out corruption, creating a level playing field even as national companies tendered low bids that increased competition. Prices for hauling trash didn't drop so much as they plummeted. The same company whose rates I had negotiated from $25,000 to $10,000 ended up paying the new norm, about $500 per month.

Having survived all these challenges, Great Forest began to grow faster than I'd ever imagined possible. We blossomed because of and commensurate with the rise of corporate sustainability. Throughout the late '90s and into the new millennium, we set up recycling programs for approximately two out of every three buildings in New York City, as well as cities like Philadelphia, Washington, White Plains, and more. Our client base became larger, so again we added services.

From its humble beginnings purveying recycled paper, Great Forest now offers consulting in all aspects of sustainability, including recycling and efficiency, to large clients like Vornado Realty, Boston Properties, Cushman & Wakefield, SL Green, Jones Lang LaSalle, HSBC, Citibank, Goldman Sachs—you name it. We've become something of a fixture in Manhattan's powerful real estate industry. At the writing of this book, we've expanded to a staff of over forty people working out of our Fifth Avenue headquarters in Harlem. We've never really advertised, nor will we ever at this stage, I bet. We really have no reason to. By this point, most corporations in the eastern United States know us as go-to guys in sustainability work.

After a turbulent decade, I suddenly found myself with the wherewithal to buy a nice old carriage house on a lake up the Hudson River, in Tuxedo Park. The place had its own dock and a deck and a big open fireplace, and I loved it. My life took on a slightly Gatsby-esque cast. I found myself throwing nice parties, joining a country club, and practicing my golf swing in earnest. I also found myself smiling a lot for no good reason at all. Why not? It was easy to congratulate myself. Hadn't I gone from rags to riches in little more than a decade? I had money. I had a wonderful family. I was living the American dream. I had built my business from nothing and made it in the toughest city in the world.

But Aesop's famous quote kept coming back to haunt me. "I thought these grapes were ripe, but I see now they are quite sour." One day, I woke up and looked around at everything I had. Rather than feeling content or magnanimous, I felt disgruntled. The situation I found myself in was exactly the one I had feared would occur if I'd stayed with IBM.

While material success is a lovely thing, and I highly recommend it, I still felt a certain emptiness, as though I hadn't found the proper yardstick to measure my life against. To paraphrase my namesake, Buckminster Fuller, what is the point of living if your life is not well-lived? This question has always haunted me.

I will tell you quite candidly that I have no particular religious convictions. I believe that when I die, that's it; my essence is finished, my time is done. To my way of thinking, it seems that to *not* do the most with the time I've been given on earth is a waste, not to mention awfully boring, as well. By my best reckoning, I had another forty or fifty years left to me at that point. Why not make the most of it? What could I do that would really make a difference?

The next day, I initiated an experiment I called VTank, short for "virtual think tank." Using an email thread, I reached out to the smartest people I knew around the world, connected them, and instigated an online discussion. My goal was to generate an answer to a single, simple question: What does the world need right now?

"No idea will be deemed too outlandish or strange," I wrote. "Let's keep our minds open, good Ladies and Gents. We will cast our net wide, let its weights sink deep, then haul it back in to shoreline to see what fish, if any, we've caught."

That email thread bounced around for several months as our inquiry raged. We discussed such topics as the use of contraception in developing nations, nuclear nonproliferation, population explosion, and the scourge of guns in the hands of children. Passions flared on several occasions, but always to prove a point or champion a cause. Whenever disputes arose—and they did—we simply returned to our core question: What does the world need right now? The level of discourse excited me, as did my VTankers' willingness to share their knowledge, feelings, and experience.

By the end of our process, we'd generated a list of problems that we felt required attention. So I hosted a couple of weekend retreats at my house in Tuxedo Park. Eight or nine people attended each session. If any of this sounds heady or formal, I assure you, our meetings were anything but. In fact, our process was deliberately eased by liberal quantities of good wine and hickory-smoked barbecue. The issues we'd chosen to tackle were weighty enough; I wanted the discussions to be engaged and to have fun. All ideas were on the table, but I asked everyone to focus on the environment.

In the evening, we gathered in my living room to discuss the best ideas. Participants called out their findings while I jotted them down on a whiteboard I'd set up. Where in the environmental sphere, I asked, should we train our efforts? What specifically should we endeavor to fix?

At the end of our final session, I stood by the whiteboard with my friend Patricia Cassidy. We had divided the whiteboard into four quadrants to represent the environmental space, sort of in the manner of a Punnett square. In the left-hand squares, top and bottom, we wrote GREEN and BROWN to represent the green and brown agendas, respectively. In the right-hand squares (again, top and bottom), we wrote WEST and SOUTH, with WEST standing for the rich countries and SOUTH indicating the deep slate of the poorer countries on earth.

"The green agenda is well-looked after in the West," noted Mary, an old friend.

Almost everyone agreed with this. We could all cite dozens of organizations that were engaged in powerful, transformative work. Save the Whales. The Wildlife Conservation Society. The National Audubon Society. The World Wildlife Federation. Conservation International.

"Look at the Audubon Society," said my friend Peter Hosking (aka "Oz"). "In the West, we pay more attention to birds than practically anyone in the South can imagine."

"Right, and it's still not enough," someone grumbled.

Oz held up his palms. "I agree," he said. "But how big is the Audubon Society's budget? A hundred million dollars a year? Maybe more? Let me be clear, I support Audubon's mission a hundred percent. But if our goal here is to pinpoint underserved sectors . . . "

When I looked from face to face, I could see that Oz's point had hit home. After a bit more conversation, we concluded that the GREEN box in the WEST column was already well-represented. And so, using my marker, I drew a big X through it.

"Okay," I said. "What about green-South?"

We all agreed that most of the leading Western green

organizations maintained offices in the South. In all fairness, these offices were plagued by underfunding and the sheer immensity of the challenges they faced. But they were still present, still active, and—in all likelihood—still in better straits than not. Which is why, in almost no time flat, I drew another big X in this square.

"So much for green," I said. "That leaves the brown column, with slots for both the West and South."

More discussion followed. The West, we agreed, had its share of pollution problems. But the West had stop-gaps already in place: Superfund regulations, environmental enforcement agencies, and legions of lawyers demanding compliance.

"I suppose we could make some kind of impact in the brown-West sector," I ventured.

No one disagreed. But then I tapped my marker on the empty Brown-South box.

"Who's looking after the brown problem in developing nations?" I asked.

Dead silence. Everyone wracked their brains, but after a couple of seemingly endless minutes, no one could name a single group, governmental or otherwise, who had made that issue their purview.

I drew a big X through the Brown-West box and rapped on the Brown-South one with my knuckles.

"This space is a big fat zero," I said. "We'll make our impact here."

CHAPTER FOUR

Brown Is Just as Important as Green

By the end of that week, I had set aside some office space at the Great Forest headquarters for my new endeavor. A few days later, I hired a bright young woman named Sara Kate Gillingham to serve as the venture's first worker bee. It felt like a damn good start, but the odds were deeply stacked against us. Speaking plainly, we had no idea what we were doing. We were like the deaf, blind, and mute leading the blind and, once again, I found myself painfully aware of the need to educate myself.

"First matter of business," I said to Sara Kate. "We need to contact as many people as possible. Anyone and everyone in the field who can share their wisdom. Let's aggregate what they tell us and try to concoct a set of early working protocols."

I soon learned that a sort of standard model exists for international organizations hoping to branch into new territories. First, you hire a Western expat who lives in the region you want to do business in. As an alternative, you can fly someone in who's previously worked in development, and embed him or her. Next, you set your agent up with a Western-level salary, something respectable—say, $120,000 a year. You buy them a house and rent them an office where they can greet dignitaries and potential supporters. You pay for their coffee boy, their car, and their car's driver. You pad them with an expense account. Health insurance. Retirement. Paid vacation time, travel included.

You do all these things, hoping the expat will earn his or her keep. Specifically, you want them to leverage their knowledge of the country's culture, language, and laws to pave inroads for your organization. I nixed this idea right away. For one thing, it was too expensive. Going the expat route demands about $250,000 just to get off the ground, and many people warned me that it could take years before things started running smoothly.

One person I talked to got right down to brass tacks.

"A lot of organizations spend millions before they sink their teeth into their first project," he said.

Not mine, I thought.

At that point, I was funding my new venture entirely out of Great Forest to the tune of about $100,000 a year. I wanted to make a difference, sure, but I could not bankrupt myself on sheer speculation that, one day, we might get our act together. From a business perspective, that didn't make sense to me.

But something else turned me off from the expat model. The more I thought about it, I didn't feel comfortable paying Westerners to give advice to people in the South. That notion struck me as patronizing and counterproductive. It still does. To me, it implies that people in the South—the people we hope to help—are somehow inferior, aren't intelligent, don't know their own culture, and have no investment in effecting change for their homeland.

There's got to be another way, I thought. Another model that works.

After a bit more research, we found that we could hire local staff in the countries where we hoped to work. These would be staff with master's degrees or PhDs who would work initially as part-time consultants, then progress to full-time work once the office had started to gather some steam. As local people, they would be much more effective than expats. They would have skin in the game, after all. This was their country, and these were their people, their problems. Likely, they would understand the issues being confronted better than Westerners ever could, as well as the proper (and sometimes delightfully improper) channels for getting things done. People would not only trust them because they were locals, but also because they were luminaries, accredited academics trained by local universities. And, thanks to drastic differences in currency values, we could usually hire one of these experts for just $8,000 to $10,000 a year.

"Our experts will be thrilled to earn that kind of money," Sara Kate noted. "Ten thousand dollars American is a fortune in most of the countries we hope to do business in. And here's another bonus. Presumably, those salaries will buy goods and services that fuel local markets and raise the overall standard of living. That's economics 101."

"So that's the model we'll use," I said.

Most of the people we consulted issued the same piece of advice: don't disperse your efforts. If you really want to attack the problem of global pollution, concentrate on one country, they said.

But which one?

To help us make our decision, we developed a methodology for analyzing a nation's pollution profile and quantifying it on a form. We asked ourselves questions like:

- Does the country have any critical, compelling, and hazardous pollution problems?
- Does this country have good legal structures, meaning laws that presumably could stop people from hurting each other via pollution?
- Has this country established agencies to enforce those laws, and what kind of job are those agencies doing? Do they have any teeth?
- Does the country have NGOs that could serve as watchdogs to any campaign to effect change?

We also asked ourselves more practical, operational questions, such as:

- How does this country dispose of its solid waste? By which we meant regular old-fashioned garbage. (Most Western countries take this for granted. But simple, run-of-the-mill trash can create legions of pollution problems when it isn't handled properly.)
- Where do this country's citizens get their drinking water?
- How do they handle basic sanitation? Medical waste? Industrial waste? Hazardous waste (meaning anything toxic to humans)?
- What legacy problems exist in this country, and how are they being handled, if at all?

Once we'd developed our methodology, it was time to fill out the forms we had made. This wasn't an armchair exercise. I wanted to conduct each country review personally. My old friend, Peter Hosking—aka Oz—offered to accompany me.

"Where do you want to begin?" he asked.

Our original list of venues included larger nations such as China, India, and the Philippines. After some debate, however, we decided not to tilt at windmills. Starting out, we'd focus on smaller countries whose systems would be easier to investigate and where our travel funds could enjoy more stretch.

"It'll help if we choose a place where English is spoken," I said. "That will negate any need for interpreters."

Oz checked the list of locations we'd generated.

"Mozambique, Tanzania, and Cambodia," he said. "Hmmm. Which one shall we visit first?"

He kept going down the list and tapped an entry with his fingertip.

"Thailand," he said. "It's perfect. You and I have been there before, remember? Back in college?"

Of course I remembered. When you grow up in Australia, everyone visits Thailand; our countries are practically neighbors.

"We got along fine there years ago," Oz said. "Plenty of people speak English. And . . . how do I put this delicately? As I recall, the place could use a little help cleaning up."

"Thailand it is," I said.

By the end of the day, we had booked our flights and made appointments with a doctor who specialized in infectious diseases to update all our shots.

◆

Some people call Bangkok the Venice of the Orient because it's a spiderweb of *khlongs*, narrow canals used to transport goods, people, and waste by boat. But for all the convenience they provide, khlongs can be filthy. The locals bathe in them, drink from them, and eat fish from them while dumping their trash in them and using them as toilets.

In one particularly poor district, Oz and I walked past a group of canals choked solid with sewage, garbage, and floating debris. Fed by rancid water, the inhabitants of a nearby slum were plagued by disease and malnutrition. The place was infested with insects; flies and mosquitoes had practically taken up residency.

Oz looked visibly shaken by the conditions we saw.

"You know," he said. "I know we're just here to survey this place. But while we're at it, couldn't we do something to help these people out?"

He was right. We had to do something.

Back at our hotel, I made some calls and set up some meetings. Over the next couple days, Oz and I hopped from one government office to the next, trying to make headway through the proper channels. Frankly, however, we stood no chance, not in just a few days.

As near as I could tell, Thailand's bureaucrats spent their work days lounging in comfortable, air-conditioned suites at the state environmental division. They sipped tea, and everyone we met wore nice, new clothes and good shoes. They had fresh haircuts and were well-dressed. They would repeatedly talk about what they called a "landmark" white paper, which the government had written several years previous.

"A white paper," I repeated.

"Yes, yes!" said one of the local officials we met. "On the subject of local pollutants. It's very well-conducted, this paper. Very thorough. The Germans helped us with it."

"But the canals are still clogged," Oz said.

"Ah," said the official. "Yes, that's true. A very sad situation. But this matter, you see. It is currently undergoing strategic analysis."

"That's great," I said. "Now how about sending someone down to the river to pick up the trash?"

"Oh no," the official said. "No, that won't do! There are protocols we have to follow. Statutes, sponsorship, barriers, studies, and blah blah blah blah legislation. You see? And blah blah blah *research* that must be performed. Consultations, many of them. With experts. We have to hire and consult with the experts first. And then hold some meetings. Quite a few, in fact, before we can proceed."

"Look, mate," I said. "How long do you think all that will take?"

"We are moving very quickly," he said. "By next year, we should understand this problem better."

Oz and I looked at each other.

"Well, that seems really clear," Oz said.

"Very much so," I said. "Thanks so much for your time."

We left that office, returned to the canal district, and found a group of local men who owned a barge. Negotiations ensued, at the end of which, I wrote them a check for a few thousand dollars, enough money to employ those men for a few months. They set to work immediately, skimming the trash off the surface of the canal and hauling it off to a garbage dump.

Oz and I extended our trip to stick around and watch them work. What we saw amazed us. So much garbage got picked up in a matter of just weeks. Free of obstruction, the water started flowing again. There were fewer insects. The place was cleaner than it had been in years. Sure, we had not solved the source of the garbage. And if we stopped skimming the trash, the place would soon revert to its former mess. But it was something. It was a start.

This simple act of skimming the canals impacted tens of thousands lives, at least temporarily. We hoped our efforts would be picked up as a model by the local authorities. That happened eventually, but not for almost another decade, which was a lesson in perseverance. That one little project stirred up some action and responses, not all of them positive, but something got done for the first time in ages. This offered its own kind of satisfaction.

"Isn't this ironic?" said Oz. "You funded local people at local rates and look how quickly the job got done."

For the rest of that year, we continued conducting our country reviews. In every location we visited, we commissioned a tiny project or two, and always to good effect. There were challenges, of course, but that was always part of the fun. To my mind, a project without a challenge is worthless, a gift you receive that you don't really want. The change we were making took hold of me like a drug. We were empowering people to take control of their lives—and in many cases, to *retake* control. And honestly, what could be better than that?

In Phnom Penh, we found mounds of trash discarded by tourists choking a beautiful historic park. Our brief investigation determined a simple cause: the local municipal government didn't have funds to purchase trash cans, or hire staff. So I wrote another check for $10,000, enough money to purchase the cans and hire some neighbors to pick up the trash while funding a local nonprofit that would oversee the work.

Moving forward, this became a signature facet of our methodology. We would generally hire a local partner—sometimes the local university, other times an NGO who was good at managing such projects—to make sure that the work we commissioned got done properly. Also, in those very early days, Oz would act as a sort of country coordinator. From time to time, he would fly to the sites of our various small projects. His job was to manage any issues that arose.

"People are people," Oz said to me once. "The majority of us want the same basic things, right? Food, clothing, shelter—the obvious needs. But people also crave more intangible things, like opportunity. A chance to prove themselves. A chance to raise their children safely. Pride in the place they live in. And so on."

I agreed then, and I still agree now. By simply removing the barrier to these wants, people will, by and large, take any ball you give them and run with it.

Over the years, a lot of people I've worked with have commented on how much I trust people. They're half right, I suppose. What I really trust people will do, nine times out of ten, is act in their enlightened self-interest. Even at this earliest stage of my career, it struck me as obvious: we don't have to scale new heights of humanity to combat the brown problem. If anything, we just have to level the playing field. The people we're helping will do the rest.

While touring Cambodia, Oz and I were shocked to learn that the country had almost no environmental legislation. This struck me as absurd. How can anyone expect bad actions and actors to cease when there aren't any laws to prohibit them?

Once we arrived back home, I began researching people I thought could help. After a bit of digging, I contacted Dr. Richard Stewart, professor of environmentalism at New York University. Dr. Stewart is an expert on best practices and international guidelines for sustainability. Once I acquainted him with the problem, he and some of his students reviewed legislation for more than a dozen countries similar to Cambodia in size, economics, and governmental structure. He then drafted Cambodia's Hazardous Waste, Clean Water, Clean Air Act, which we had translated into Khmer and sent to a contact I'd met during my trip: Mok Moreth, Cambodia's minister for the environment.

Basically, our efforts replayed the same technique I had used in Brazil: do the work yourself and slip it into the hands of some influential person. It's positively amazing how much you can get done in life when you don't care who gets the credit. Mok Moreth was very happy to be handed a piece of well-researched draft legislation. It was another win-win outcome obtained for a relatively meager sum. As an organization, we'd also stuck a very bright feather in our cap. How many small nonprofits can claim to have drafted such a body of environmental legislation for an entire country?

More projects followed swiftly. In Tanzania, we saw a river with water that ran neon blue thanks to chemical waste dumped into it by a local toothpaste factory. The people living downstream from the plant suffered annoying skin irritations, but this turned out to be another simple fix. I got an appointment with the head of the plant and took a seat in front of his wide wooden desk. We spent half an hour making small talk about politics, sports, and the personal lives of American movie stars.

"What is the draw to American football?" asked the plant's head. "I find cricket much more interesting, don't you? Now tell me all about Brangelina. Have they adopted any more children? Frankly, their actions confuse me. Why would Ms. Jolie give birth in Namibia when we could make her much more comfortable right here in Tanzania?"

At what felt like the appropriate time, I steered our conversation toward the river and its condition. The water downstream, I noted, bore an uncanny resemblance to a certain local brand of toothpaste that the plant, it just so happened, produced.

"How odd that a river has turned blue," I said. "Is there any possibility that your filtration units have malfunctioned?"

The head of the plant stared at me for an awkward beat. Then he vowed to check on the matter. Within hours, the water was running a normal color again, which made me want to throw up with disgust.

He'd known what was going on, of course. He just didn't care. We later learned that his filtration units hadn't malfunctioned; he'd switched them off to save money on power. When people complained, he ignored them, but not me. Why? Because I was a Westerner and said I worked for an environmental nonprofit. A Westerner who might pick up the phone and call up my wealthy Westerner friends, and they might be appalled to hear some story about a river as blue as a glistening gobbet of extra-strength cavity-fighting formula. Government liaisons would come snooping. The media might send cameramen. So the man threw a switch and voilà! The river ran clean, the people's skin irritations abated, and the problem was solved.

Or was it?

A bit later on, we got wind that this plant was up to its old shenanigans, so we had to exert a bit more pressure. This time I funded a more able local nonprofit to serve as the factory's watchdog. Live and learn.

After a number of these types of jobs, we decided we needed a name for the kind of work we were doing. But which name would best fit an organization like ours? To help me decide, I reviewed what I thought was the mission of my fledgling nonprofit.

Of all the NGOs working for the public good—some big, some small, some politically well-connected, some very well-heeled—ours was the only one I knew of that would be tackling pollution problems in low-to-middle-income countries. Specifically, we would fill the crucial gap between local communities and the government or non-government agencies that could offer the most resources. In other words, we would handle the problems that nobody else wanted to touch. We would employ scientists, public health experts, and environmental engineers who would put boots on the ground and dirty their hands in some of the worst hellholes on the planet. And our people would get results.

What kind of name would signify that?

We thought for a while and finally settled on a name: Blacksmith Institute.

Strange name, people said, for a group that deals with human health and the environment.

Maybe. But the more I thought about it, the more this image stuck with me: a bearded, soot-smeared, sweaty-backed hulk towering over a deep forge filled with hissing, hell-hot coke. In my mind's eye, I could read the scars on the backs of his hands as clearly as if they were newsprint. I saw the patches where his skin had been calloused by sparks, made rough and scabbed by the grips of his tongs, scorched by hot iron and boiling smoke. This guy hadn't sculpted his muscles in a gym, he'd earned them from long days spent pounding his anvil, grunting in counterpoint to the musical ping of his hammer. He was ugly and he was intimidating, but his gaze was serene, and filled with the craftsman's inner peace—the satisfaction that only comes from knowing you've done good work. Because, at the end of each day, this man could hold up a horseshoe, a plowshare, a link in a chain. His creations were rough, but sturdy as hell. He'd built them to work and to last.

Here, says a blacksmith. I made this for you. It's not pretty, but see how it gets the job done? Beware of those who offer you pretties that don't.

Flash forward a decade. With some great projects completed with resounding success, we determined that we needed another name. "Blacksmith" required too much explanation. As powerful an image as it was and is, it doesn't immediately conjure up the image of a clean and safe world. So we went back to the drawing board and rebranded ourselves as Pure Earth. We held some workshops and plenty of discussions to vet the new name. In the end, it was Dev Patel, the brilliant actor who played the hero in *Slumdog Millionaire*, who pushed us to choose Pure Earth. He and actress Freida Pinto are passionate supporters of the work against pollution. In two words, our new name tells the story of what we're hoping to accomplish through our work, and it's what we'll use moving forward. (Well, almost. A lot of people still love the name Blacksmith Institute.)

If our new name feels "prettier," it doesn't change how we attack pollution. We can't clean up the world without getting our shoes muddy. I will never dispute the value in studying a problem—what I call "the white paper phase." But once we've established what needs to be done, it's time to cordon off landfills, break out the shovels, and crank up the dozers. An action must follow a thought, or all good thinking comes to naught.

"If we act, then change will begin to occur," I said. "That's the good news."

"There's bad news, too, I'm afraid," Oz said.

He picked up the list we'd begun to compile, the one that held the names of the most toxic sites on the planet, and waved it under my nose.

"There are more of these sites out there than we ever imagined," he said. "They're bloody deadly and their numbers keep multiplying. A couple hundred million people are affected each year."

The jobs we were handling started to grow in size and complexity, which pleased me. It's undeniably helpful to install trash cans in parks and so on. In my mind's eye, however, I envisioned Pure Earth handling more truly hazardous pollutants, and on a much larger scale.

Pure Earth grew by leaps and bounds. By 2007, we had a solid board of directors in place whose members sought new and provocative ways to raise public awareness of our mission. One of the best ideas came from Josh Ginsberg, who worked in the international program at the Bronx Zoo, the Wildlife Conservation Society. Josh's brainchild cut right to the chase.

"People really get behind top ten lists," he said. "They worked pretty well for David Letterman, right? Because it's mind candy—information that's simple enough to activate people on the left side of the bell curve, and clever enough so people on the right side remember what you're telling them. Why don't we do a top ten list of the world's most polluted sites?"

So that's what we did. The campaign took off and we never looked back. In fact, the idea worked so well, we recycled its basic template for the material you are about to read. It is important to note that the problem of global pollution, unfortunately, is larger and deeper than even this varied sample suggests. But this seems to be a good enough place to start.

Join me now on a tour of our polluted world. It might be a world you didn't know existed. Frankly, I sort of hope that's the case, since I'd like more people to open their eyes and possibly involve themselves in this cause, which I've come to care deeply about. I'd like your help in making a difference. Together, I've already seen that we can.

You might be surprised by how much of an impact you can create, and with laughably little effort.

Before we move on to our tour, however, I want to acquaint you with a few simple truths I've discovered that correspond to the pollution problem worldwide. These axioms have reaffirmed themselves often and in many diverse locations over the years. I offer them now to provide a foundation for your understanding as we take a deeper, more specific look at the top ten pollutants threatening planet earth.

AXIOM #1: Pollution is not an inevitable result of development.

In fact, the opposite is true: brown problems hold development back. The most common argument I hear along these misguided lines is what I've come to call the Industrial Revolution Comparison, which goes like this:

Nations like China and India are polluted because they've become industrialized and have grown without caring at all about their environment (caveat: this is generally true). Also, mirroring the West's Industrial Revolution, this type of rapacious growth is the only way for rising nations to develop themselves effectively (this is the piece of the argument that is quite wrong). Pollution is inevitable, and is also something that must be fixed once the country in question improves its economic circumstances.

This argument has two main flaws. First, it ignores the fact that newer technologies are both cleaner and more productive than their earlier counterparts. Investment in newer technologies is more often justified by increases in output rather than regulatory compliance. There is, therefore, every reason to hope that the poorer countries on earth will invest in the latest technologies and leapfrog the problems that beset their richer counterparts. And that this, in turn, will require them to not spend so much on cleaning up pollution.

The second flaw: it is now quite clear that pollution hampers economic growth. A sick population is not a productive one. Cities that are clean and green grow quicker than dirty ones; the latest research is very clear on this point. None of this was understood back during the Industrial Revolution, and the West had no choice but to stumble its way through the stages it did. But developing countries need not do so. They can choose to avoid these mistakes from the start.

Along these lines, government incentives that allow only

clean industries to operate will actually benefit economic growth, and should be the standard. Sadly, governments have not understood this point too well. Overall they have lacked the willingness and capability to require pollution-minimal industrial growth.

Axiom #2:
Local governments generally do not possess the expertise or the resources to manage their pollution problems.

Some people take advantage of minimal or nonexistent sustainability standards to rape a region, poison its people, and slip back over the nearest border with suitcases full of cash. Countless times, Pure Earth has arrived at the scene of some truly awful pollution and found that someone knowingly contaminated the region to maximize profit. Worse yet, they left behind their toxic messes for somebody else to clean up, or not. Such instances can and must be addressed through the proper legislative and law enforcement channels. This is easy to do in the West, but sadly almost impossible to accomplish in the South under current conditions. Which leads me to Axiom 3.

Axiom #3:
The West, having already done so much to handle pollution, can and should make a difference in the South.

There are several arguments for why the developed world should pay attention to pollution problems in the global South. First, much of the growth behind the explosion of pollution in poorer countries is a result of the richer countries pushing their dirty industries overseas. An inadequately educated government, perhaps with corruption issues, is easy to sway with the promise of jobs and taxes, even if it is not to their benefit. And while we want to buy the latest toys and gewgaws, we should not do so at the cost of other human beings. Since Western demand is a big part of this problem, we have a moral responsibility to help.

Second, a safe world is one where poverty is low. Poorer countries are less educated, filled with people without hope. They are also more likely to rise up and overthrow, or turn into guerilla armies, or listen to crackpots that think they should attack Western shores.

A country with a comfortable middle class is less likely to become a breeding ground for terrorists. The more we can do to help other countries out of their poverty spiral, the safer our world will be. And yes, this includes pollution, which is often the clearest indicator of inequalities in a country, and thus an incentive for action.

AXIOM #4: Fixing brown problems needn't be expensive.

This might be the most important notion of all. In most of the projects we work on, the cost for saving a human life can be as low as forty to fifty dollars per person—less than most people would pay for a single meal in a good restaurant, or a two-pound bag of coffee.

I've noticed that many people working on pollution in the West meet this axiom with suspicion. Surely, they think, it costs billions to clean up pollution. Why, look at what it's costing to fix the PCBs in the Hudson River!

To examples such as these, I say: yes. Cleanups in the West can be expensive, but this is only because we are in the expensive phase of the cleanup process. The earlier projects we handled back in the '50s and '60s were cheap by comparison, and very effective at saving lives. Once the low-hanging fruit is dealt with, then the fixes become pricier. The developing world, however, stands at the beginning of the cleanup process. For this reason, helping them clear up their pollution problems doesn't cost as much as you might think.

Axiom #5:
Most toxic sites are the byproduct of local companies, not big Fortune 500 companies.

It doesn't matter if we're talking about the Soviets in Horlivka, small-scale miners in Indonesia, or the power industry in Delhi, India. A basic premise underlies each situation: environmental contamination flourishes in places where the welfare of other human beings has not been properly considered. Governments have not had the resources or the incentives to act. Local factories have not had guidance or controls.

Small operators often do not know of the dangers of their operations. But when the multinationals of the world enter a low- or middle-income country, invariably they import Western standards with their operations. Why? Chiefly because they want to avoid NGO criticism, which detracts from their reputation.

But the big companies' operations tend to be so good that local firms begin to emulate them. So while it may feel revolutionary to rail at big companies for exploiting developing nations, this is often inaccurate. Very likely the same companies being vilified are actually leading communities in their host countries to better ways of life. Obviously, some exceptions exist, but these are remarkable because of their infrequency.

For example, as of the writing of this book (late 2014), Pure Earth's database of severely polluted sites includes over 3,000 locations, and this number is growing fast. The criteria for being included on this list are compelling. A known human toxicant must be present in a pathway of exposure to local people (not just workers) at levels that significantly exceed standards set by organizations like the U.S. EPA. The number of people exposed must also be non-trivial, numbering at least in the thousands. In other words, these sites all represent very, very bad instances of pollution. By the way, the professionals who evaluate sites for inclusion in the database are incentivized simply to find pollution; they pay no attention to the polluter.

In reviewing the database, we found that only two out of 3,400 polluted sites have multinationals involved. To reiterate: the overwhelming majority of acutely polluted places on earth are the result of local industry and mining, legacy contamination, and small-scale activities—not the big guys everyone likes to blame. The bottom line is that we are not going to solve this problem by blaming multinationals.

Okay, that's enough education for now. Let's look at some examples of the top ten forms of pollution affecting our planet right now.

PART TWO

The Top 10 Pollutants Threatening Planet Earth

CHAPTER FIVE

Cities of the Damned

Back in 1895, the famous explorer Frederick Russell Burnham took a job with Cecil Rhodes's British South Africa Company. Rhodes dispatched Burnham to look for mineral deposits in a broad lump of land some 10,000 miles square that wrapped Central Africa from east to west, like a corset. He was also interested in ways to improve and capitalize on river navigation throughout the region. In the course of his travels, Burnham stumbled across massive copper fields, which he described like this in a memo to Rhodes:

> About 200 miles north of the [Victoria] Falls on the Incalla river, and twelve miles from the Kafukwe [now the Kafue River] and still on the high plateau is probably one of the greatest copper fields on the continent. The natives have worked this ore for ages, as can be seen by their old dumps, and they work it to-day. The field is very extensive, and reaches away to Katanga [the southernmost province of Congo] . . . The natives inhabiting this part of the country are skilled workmen, and have traded their handiwork with all comers, even as far afield as the Portugese [sic] of the West Coast and the Arabs of the East. These natives, being miners and workers of

copper and iron, and being permanently located in the ground, would give the very element needed in developing these fields.

Few scouts of that era would have recognized the copper fields for what they were. As it happened, Burnham had an extensive history with the metal and the mining methods used to extract it. He was an American, and before leaving the U.S. to work in Africa, he had earned his living in various professions throughout Mexico and the American Southwest, including several stints as a prospector.

In subsequent letters, Burnham noted that, based on his experience at mines in Arizona and Montana, copper was potentially more profitable than gold and silver. But even he couldn't comprehend the full potential of what he was looking at. Beyond copper, the fields held vast strains of cobalt, lead, lithium, emeralds, and lesser-known substances such as carrollite, Bornite, and Pyrite. In other words, Burnham had discovered a windfall of resource wealth in an age when industrialization was sweeping the globe and placing these same resources in high demand. The colonials nicknamed this cornucopia region the Copperbelt. It would change the course of history.

Eager to turn a profit, foreign companies swooped into Central Africa and snatched up the region's mineral rights. They built smelters, factories, work camps, and railroads, hiring Africans to do all the dangerous work in exchange for European pittances. Trees groaned and crashed to the dirt. Mines raked the countryside, scraping it clean. Hot soot billowed from smokestacks and vent flues, rising in columns to blacken the sky.

Most of present-day Zambia sat smack in the middle of all this excitement. In particular, the settlement now known as Kabwe was sitting on massive deposits of lead that ran an astonishing twenty percent pure. The smelter at Kabwe went into operation in 1902. For decades, the mines in and around that region gushed zinc, manganese, silver, vanadium, and cadmium, but mostly lead. Immensely successful, the Kabwe smelter became the most profitable facility of its kind on the continent. It provided jobs for the local population at some of the highest wages in the region.

The Copperbelt's industrial boom rippled out, creating unprecedented prosperity throughout Central Africa which, in turn, became a key support strut in Zambia's declaration of independence from Great Britain in 1964. In October of that year, Kenneth Kaunda, a former schoolteacher turned socialist leader, became the first president of the newly independent Republic of Zambia. However, after brutally crushing a small-scale civil war, Kaunda became increasingly intolerant of opposition. He proclaimed his country a one-party state and, mimicking Soviet reforms, began wresting pivotal economic resources from the hands of foreign-controlled companies by acquiring majority stakes in them. In other words, Kaunda meant to kick out or subjugate the large-scale, colonial-based industries that, by tradition, had dominated Zambia's economy since the days of Cecil Rhodes.

Kaunda targeted banking concerns, insurance groups, building societies, and—not incidentally, nor without irony—mining companies. One such firm was Anglo American, perhaps the world's largest and most diversified natural resource group and the majority stakeholder in the DeBeers diamond empire. Along with fellow resource giant Rhodesian Selection Trust, Anglo American's holdings were absorbed into a state trust called Zambia Consolidated Copper Mines, Ltd., or ZCCM, which Kaunda ultimately directed.

Kaunda was playing a dangerous game, keeping the United States and Europe at arm's length while forging alliances with anyone who seemed useful to him. Apart from the Soviets, this list included Egypt's Gamal Abdel Nasser and Saddam Hussein of Iraq. Domestically, he pursued a policy of paternalistic socialism, which included a vigorous welfare state paid for by mining. The potent assets held by ZCCM led Zambia to experience yet another economic boom. By the early 1970s, she was the most highly urbanized nation in the sub-Saharan region. Fully ninety-five percent of Zambia's exports centered on copper and related minerals. Zambia's citizens enjoyed the highest standard of living on the continent. The country was hailed as a bellwether for the New Africa. At which point, catastrophe struck.

In 1974, the Gulf States levied embargoes in reprisal for U.S. support of Israel during the Yom Kippur War. Oil prices skyrocketed while mineral prices tanked, taking Zambia's prospects with them. To complicate matters, the same state entities that oversaw Zambia's nationalized resource sector hadn't properly managed the country's wealth. Less politely, one could say that corruption and cronyism had bled the national coffers dry. Kaunda had succumbed to the same delusions of grandeur and indispensability that had made tyrants of nominally democratic African leaders like Julius Nyerere of Tanzania and Robert Mugabe of Zimbabwe. Pro-Kaunda forces intimidated and jailed opponent politicians. They shut down newspapers and strangled free thinking, even while their country was going to hell all around them.

Within months of the Gulf State embargoes, Zambia's treasury had expended its foreign exchange reserves. Her storybook rise reversed itself. The country soon found itself utterly broke and crushed under relentlessly mounting debt. Once a herald of the "New Africa," Zambia fell into a grim cycle of debt, dependency, and political chaos. Exacerbating matters, the ZCCM and other state agencies proved remarkably incompetent at managing the country's wealth. By the mid-1980s, Zambia was broke and deeply indebted to the International Monetary Fund, which called for extreme measures to stabilize the nation's economy.

Specifically, the IMF wanted Kaunda to reduce his country's dependence on copper and mineral trading; to devalue Zambia's currency, the *kwacha*; to slash government spending; and to eliminate food and farm subsidies. Prices in Zambia hit the roof. Riots broke out in the capital city of Lusaka. Desperate to stay in power, Kaunda broke with the IMF and agreed to privatize the very same industries he had worked so hard to nationalize nearly two decades before.

To prop up its collapsing economy, the Zambian government borrowed money from other nations. This only made matters worse. When Zambia could not repay these loans, its sinkhole of debt plunged deeper until, finally, the International Monetary Fund and the World Bank stepped in. The banks offered emergency funding, albeit tied to strict reforms. Sadly, they were too late. The damage, by that point, was already done.

Over the next fifteen years, the Copperbelt, once the source of Zambia's wealth, languished in a state of near total dormancy. The Kabwe smelter and almost every other facility like it shut down. No one could afford to keep them running. Even in the best of times, these enterprises had never been well-maintained. In order to maximize profits, the ZCCM had cut almost every conceivable corner. Safety features didn't exist. Technology upgrades were scoffed at as superfluous. Operations had been slashed to the bone. Crepitating to begin with, these facilities, when completely abandoned, deteriorated with shocking speed. Once a bustling industrial hub, the region became a graveyard of rotting equipment and factories.

With no mining industry to sustain them, towns in the Copperbelt atrophied into slums. Streets stood empty and houses deserted. Eerie silence hung in the air. Shops closed up by midday or simply refused to open at all. Unable to afford fuel, citizens ravaged the local forests, cutting down trees to feed their cook fires. Year after year ground by like this.

"We are dying here," said a local reporter in 2000.

He wasn't exaggerating. Without money to pay for decent food, people fell prey to chronic malnutrition and disease. Those who sustained serious injury were essentially sentenced to death since no one could afford proper emergency medical care. Children stopped going to school in order to help provide for their families. The situation was bleak indeed.

In 2002, Yves Prevost, my friend at the World Bank, had pointed me toward Zambia.

"You won't believe what's happening there," he said. "The situation is awful."

"If it's so awful, why aren't you doing anything about it?" I asked.

He snorted. "Politics! This place I'm talking about, it's one of the worst cases of legacy contamination I've ever seen. But due to the state of affairs with the Zambian government, I'm afraid our hands are tied."

Pure Earth's pilot programs in Thailand, Cambodia, Tanzania, and Mozambique had gone well. At that point, we were ready to grow. Intrigued, I purchased a ticket and flew to Zambia's capital city, Lusaka. From there, I hired a guide to show me the Copperbelt, which, at that point, had languished for nearly two decades.

The damage I witnessed was staggering. In a town called Kafue, heavy rains had ruptured a dam built to contain a reservoir of mine tailings. Also known as slimes, slickens, or leach residue, tailings are the chunks of rocks and bits of dust created when industrial processes extract minerals from raw ore. They often contain substances dangerous to human and environmental health—radioactive byproducts, heavy metals such as mercury and arsenic, and so on. When the dam at Kafue burst, these substances spilled across the countryside, polluting whatever they touched.

The tailings stored in the Kafue reservoir had been particularly rich in cadmium, and that was a problem. Cadmium is a heavy metal, and brutally toxic. Exposure to it triggers early-onset atherosclerosis and hypertension, which can lead to cardiovascular disease. Hearing loss has also been proven as a side effect, and it's been linked to certain cancers, notably those that afflict the lungs and prostate gland. But the risks don't stop there. Inhaling cadmium fumes can result in pulmonary edema, pneumonitis, and death. These same risks are present when someone ingests cadmium-contaminated food or water.

One famous case of this occurred in Japan during the early twentieth century. Cadmium and other trace metals began to show up in the water and soil of the Jinzū River in Toyama Prefecture. Rice crops along the riverbanks absorbed the cadmium; people who ate the tainted rice began to develop what the Japanese call *itai-itai* disease (literally "ouch-ouch" disease). First, they developed searing pains in their spines and legs, which caused them to walk with a waddling, duck-like gait. Next, they were stricken with mad bouts of coughing, anemia, and kidney disease. Simultaneous to this, the cadmium leached strength from their bones until they became brittle enough to fracture under the barest pretense. No longer able to walk, patients stayed bedridden for a time, and then they died.

While touring the spill zone in Kafue, I noted the area's withered trees and dead crops. The local authorities slouched about, bereft of the resources or incentive to do anything about anything. Smiling, my guide pointed to patches of new foliage—sickly brown grass and anemic plants that were struggling to rise from the cracked breast of the earth.

"You see?" he said. "This isn't so bad. Already the plants come back."

It took real effort to hold my tongue. To someone untutored in heavy metals and their finicky ways, the appearance of vegetation might indeed seem proof of nature reclaiming her own. But it was premature optimism at best. The presence of flora means nothing to a region poisoned by cadmium. If anything, this is a sign of distress. Plants store cadmium like they were born to do it. That's one of the reasons why people along the Jinzū River got so sick in the first place. Water diverted to irrigation canals fed cadmium to local the rice crops, which stored the heavy metal until they were harvested. The harvesters sold the rice to people across the region. The people across the region ate the rice and ended up eating the cadmium, too.

My tour guide must have noticed the look on my face, for he laughed.

"Look," he said. "I know this might seem pretty bad. But at least it's not Kabwe."

My ears perked up. I had never heard of Kabwe before. "What's going on there?" I asked. Meaning: What could possibly be worse than the situation I was looking at right then?

When I asked my tour guide to take me to Kabwe, he refused. Noting my interest, he clammed up fast, suspecting, I think, that he'd said too much. Turns out, he had good reason to be nervous. As I soon learned, Zambian authorities were doing as much as they could to divert attention away from Kabwe and what was happening there. I pursued the issue with different people, but the more questions I posed, the less information I got.

Finally, I bought a map and saw that Kabwe wasn't far away—about 130 kilometers north of Lusaka, which was a two-hour drive up the highway. So I went to see the man slouched behind the three-legged desk of my hotel's makeshift concierge station.

"Are there any buses to Kabwe?" I asked.

The man nodded and said, "There is one bus that leaves in an hour."

The look on his face said he couldn't imagine why anyone would go there. For me, that sealed the deal. A round-trip ticket cost 43,000 *kwacha*, or about eight bucks American. I bought a ticket and took the ride north, my curiosity piqued.

Toward the end of the trip, my rattletrap bus bounced along a sun-blasted road that looped around what appeared to be a desert the color of dark gray ash. Hoping to get a better view, I pressed my cheek to the sun-heated glass. The wasteland of slag rolled out like a snapshot of an ocean roiled by typhoon winds. Hundreds of acres of dead land piled in rolling hills of grey ash and soil.

A moment later, a sign loomed up by the side of the road, its edges blurred by rippling heat. Words written large in English said Welcome to Kabwe. Under that, some vandal had spray-painted a weeping red postscript: This Is Hell.

I could spot no end to the sea. Later, I would learn that it measured approximately two and a half kilometers square, and that a great deal of this area was covered in toxic lead slag.

The people who'd run the Kabwe smelter either hadn't known about proper disposal techniques or, more likely, hadn't cared. In any event, they had spent the past ninety years dumping the smelter's effluents in their own backyard. I was looking at nearly a century's worth of legacy contamination. It was a shocking sight.

Eventually some of those gray hills parted and gave me a good look at the smelter. Its burned-out skeletal structures ranged from one to five stories in height. Each building presented a study in rust. Old I-beams poked through blasted brick. Crumbling walls exposed abandoned factory floors. A few ancient smokestacks stabbed at the clear blue sky; they looked like the pillars of some ruined temple.

A cold feeling washed over me. The rubble of the smelter seemed less architectural in nature, more like the remains of some monstrous prehistoric beast that, at some point long in the past, had crawled out into the toxic zone to die and decay, leaving petrified bones to prove it had ever existed at all.

The edge of the waste swung into view. About a kilometer away, the gray field ended abruptly. Beyond that line was green grass and trees, then fields and dirt roads meandering up to low-slung, mud brick huts that graduated to cinderblock shacks. Then gardens. Side streets. Tiny town squares. Soccer fields. Store fronts. Old cars that chugged back and forth, carting people and wares along paved boulevards.

It looked bucolic, but it wasn't at all. As I soon discovered, the town of Kabwe itself was a toxic zone. At that point, during my first visit, it had a population of some 225,000 souls and almost every single one of them suffered from serious lead poisoning.

Lead had not stayed confined to the wasteland. While the smelters of wealthier nations use filters to scrub pollutants from their exhaust, the smokestacks at Kabwe had belched leaded fumes from the moment they began operating in 1902 to the day the smelter closed in 1994. This would have been bad enough on its own, but the climate of Kabwe is hot and dry. It rarely rains there, and strong winds normally blow from the south, particularly during the winter. For over a century, these winds had picked up loads of toxic lead dust from the smelter field and dumped them all over town. This dust covered everything—buildings, yards, community squares. It settled on streets with ironically hopeful names like Independence Avenue and Freedom Way. It infiltrated the soil where farmers planted their vegetables and from which poorer citizens culled mud to fashion brick houses. And it was wickedly toxic.

Lead occurs naturally in soil at quantities less than twenty parts per million. The U.S. EPA has set safety thresholds for lead-soil content at 200 parts per million for agricultural zones and 400 parts per million for residential areas. But such distinctions meant little in Kabwe, where houses abutted farms that abutted houses abutting vegetable patches. Even so, soil readings taken near Kabwe's prime residential areas registered between 4,000 and 38,000 parts per million of lead—ten to eighty times higher than the EPA's maximum levels. Kabwe's citizens had inhaled and ingested these particles for four generations. Lead was in their air, their food, and their homes. It was in the very soil, and it was also in their water.

As luck would have it, the smelter's wastewater canal ambled out of the mining field and cut through Kabwe's Katondo Township, home to some 20,000 people, before crossing several housing districts and looping through two other townships. The water flowing through this canal was deeply polluted, and yet, having no other plumbing, the poor people of Katondo used it for bathing and washing their clothes. The waste canal was their primary source of drinking water.

In 1995, soil readings taken near the canal surpassed 250,000 parts per million of lead. The region's toxicity was off the charts, but there was more bad news to come.

During the rainy season of 2002, the canal backed up, deluging the local countryside with decades' worth of toxic slag. The following year, a local woman named Purity Mwanza spoke to a reporter from the BBC. Ms. Mwanza made her home directly beside the drainage canal. Her six-year-old daughter, Mary, had recently become very sick.

"When the [Zambian government] health people came here," said Ms. Mwanza, "they found that each and every child was poisoned by lead. The [state] health people at one time treated them all with tablets and instructed us to wash the children three times a day to get rid of the dust. But this was too much for us mothers. One day my child was walking in the soil and her feet became blistered. She also got sores on her stomach and bad pains in her stomach because of the lead."

Another Kabwe resident, Matildah Muyunda, spoke to reporters from IRIN (Integrated Regional Information Networks, a newsletter associated with the UN's Office for the Coordination of Humanitarian Affairs). Ms. Muyunda served as primary caregiver for her two grandchildren.

"My two children last month developed a terrible skin rash

and blisters which looked like chicken pox," she said. "They could not eat for two days and when I took them to the hospital, I was told it was something to do with lead poisoning, but they were not given any medication until [the rash and blisters] disappeared on their own after some two weeks."

The admission that Ms. Muyunda heard—that lead poisoning may have caused the children's ailments—was unusual. For many years, local and national authorities in Zambia denied that problems existed in the Copperbelt, mostly for political reasons. Admission could equate to culpability, which in turn could equate to liability. And liability could not be permitted at a time when the national treasury was bankrupt. Put differently, Zambia could not afford to defend itself from hundreds, possibly thousands of lawsuits levied by its own citizenry.

Also, for every incident of toxicity to which it pleaded guilty, Zambia's government placed itself on the hook to pay for cleanup costs. The IMF and World Bank had made this point clear when they extended aid to the country: Zambia must shoulder the responsibility of cleaning up toxic mines prior to accepting bids from interested buyers. The fact that no willing buyers existed was sort of a cosmic joke. To be fair, the Copperbelt still had some of the richest mineral deposits found anywhere on earth. But even at rock bottom prices, few corporations wanted to buy the region's mines since doing so meant they would have to take on the cost of remediating unprecedented levels of contamination.

So the country was locked in a stalemate. Toxic pollution continued to grow, while the Zambian government ignored or denied the problem on an epic scale. Whenever someone cornered them, officials downplayed evidence and accused the World Bank of mounting a smear campaign to drive down the asking price of their mines. In one recorded instance, government doctors showed themselves either completely incompetent or perfectly willing to lie on behalf of the administration.

After "investigating" the situation in Kabwe, they reported that the population suffered from malaria and prescribed free milk for the children.

Meanwhile, the effects of lead poisoning were clearly visible to anyone who visited the Copperbelt. They were visible to me. The people of Kabwe, who had played an integral role in Zambia's economic rise, were being left to suffer. Many Zambians sport names ripped out of Dickens or Thackeray. Names like Purity, Christine, and Edward abound. James. Herman. Justice. Prudence. Bright, cheery names deserve bright, cheery faces. In Kabwe, however, these faces turned slack-jawed and slow. Nobody showed any sparkle. Their limbs drooped, nerveless from lead poisoning.

Lead acts as an extremely potent neurotoxin. In the U.S., blood lead levels are measured in tiny units: micrograms (a millionth of a gram) per deciliter (one tenth of a liter). According to EPA standards, human beings can tolerate about five micrograms of lead per deciliter of blood. Anything over that affects us. At twenty micrograms per deciliter, for instance, our nerves become affected, resulting in slower reaction times and loss of sensation. Hearing loss can occur at thirty micrograms per deciliter; kidney damage and infertility at forty; decreased ability to make red blood cells at about fifty. When a human body absorbs 150 micrograms of lead per deciliter of blood, the very likely result is coma or death.

In children, however, lead poisoning exacts a markedly worse toll, and manifests symptoms more quickly than it does in adults. Lead can cross the placental barrier, causing premature births and reduced birth weights, genetic mutations, impaired IQs, and nervous disorders. It can stunt a child's growth—assuming the child lives long enough to grow in the first place. It can also cause problems with vision and hearing, mental retardation, focus impairment, violent or anti-social behaviors, and death. Independent medical experts who visited Kabwe found that the average level of lead in a child's blood hovered between sixty and 120 micrograms per deciliter. In some children, however, lead levels peaked at 300 micrograms per deciliter. These children most likely died later.

You could see the lead's impact just by watching some of the kids. They had droopy feet and flappy hands—a clear sign of neurological damage. Their running looked disjointed and somewhat unnatural. Though I could not speak their native language, I imagined they would communicate somewhat vacantly, since the lead had damaged their brains, impairing them intellectually. You could also see a dark band on their gums, just above their teeth. This is called a "lead line," and it's a classic sign of lead poisoning.

It was an unthinkable tragedy, a city of the damned.

Following the Copperbelt's economic collapse, a large percentage of Kabwe's citizens found themselves unemployed. With no other livelihood to sustain them, many started venturing onto the smelter's toxic campus to scavenge for chunks of leftover lead. Once cast off as inferior, these chunks now fetched decent prices in a thriving local black market.

A woman named Christine Mupika spoke to an IRIN reporter while digging through the poisoned soil of the smelter's abandoned quarries. Barefoot, she wore no protective gear. She was fifty-two years old, she said. Her husband had died some nine years before.

"Much of my income comes from coal," said Ms. Mupika. "Sometimes I can sell two bags in one day. Zinc takes a bit longer to find a customer, and it is not even profitable."

Ms. Mupika said she earned about $0.25 for a fifty-five-pound sack of zinc, and about $1.25 for the same amount of coal. Prices, however, could shift, she said, depending on how she haggled. But her ventures into the toxic field had begun to take their toll. Ms. Mupika told the IRIN reporter that, on three occasions, she'd been diagnosed with lumps in her chest which doctors attributed to lead poisoning. They advised her to stop working in the smelter fields, but Ms. Mupika scoffed at this.

"If I don't work here, then I won't feed my five children at home," she said. "How do they honestly expect me to survive if I stop mining? I do this because I have no other means."

Some scavengers I witnessed had bored primitive mine shafts directly through the gray hillocks on the smelter's campus. With no training in how to construct a decent shaft, many were buried alive when their mines caved in. Likely, they never felt a thing. As veterans of the smelter field, exposed to unprecedented levels of lead toxicity, most had probably sustained nerve damage.

The U.S. EPA and the World Health Organization set safety thresholds for the lead-soil content of industrial centers at 2,000 parts per million. In 1995, most soil samples taken near the Kabwe smelter ranged between 20,000 and 100,000 parts per million—ten to fifty times higher than maximum safety levels. A few samples spiked as high as 245,000 parts per million.

I had gone to Africa seeking a larger project in which to involve Pure Earth. For better or worse, I had found one. But what sort of change could we hope to effect in such a dire situation?

By that point, Pure Earth's playbook was solid and presented me with a means to begin. I looked around for an NGO that could watchdog any operations on the ground. When I found none that seemed up to the task, I poked around looking for a local champion, someone to lead my efforts. Then I got hit by a stroke of blind luck. A few weeks after returning from Zambia, I received an email from someone who identified herself as Em Heartsworth Mambwe Ozumba, a former employee at ZCCM.

"I have heard that you are investigating lead contamination in my country," Em Heartsworth wrote. "Your name was given to me by Mr. Brian Wilson, program manager for the International Lead Management Center. I know plenty about the lead levels in Kabwe. Are you interested in meeting?"

Brian's name was enough to pique my interest. The ILMC is an industry group, yet laudable for all that. It works to improve the image of its member companies by promoting safe, sustainable mining and ore processing techniques at sites around the globe. Curious, I called Brian, who confirmed that he and Em Heartsworth had recently been in touch.

"But you're a little off-base," Brian laughed. "Heartsworth's a he, not a she. The 'Em' is actually an initial—'M,' which stands for Mapalo, a Bemba word meaning 'blessings.'"

"Aha," I said. "Tell me more."

I learned that, in Zambia, M. Heartsworth was regarded as a kind of latter-day prince. His father was a national legend who had grown up destitute in a remote northern mining village where there was no electricity or running water. Even so, Mambwe Ozumba the Elder leveraged his considerable intelligence and work ethic to become one of the first black people to earn a university degree in Central Africa. This, in turn, allowed him to join the small, elite group of Zambian officials who, along with a young Kenneth Kaunda, played key roles in shaping the newly independent country throughout the '60s and '70s.

Mambwe Ozumba served in various government positions, including as a member of the Zambian parliament and governor of the Bank of Zambia. But he became increasingly frustrated and vocal about the reforms that Kaunda's government fostered. As reward for his outspokenness, he was jailed for treason and given a death sentence that was later repealed; the national Supreme Court ordered Mambwe Ozumba released when it was found that his "confession" had been obtained by torture. His treatment in prison did permanent damage to his health, and he died relatively young at age sixty-three in 1994. To this day, Zambians regard Mambwe Ozumba's widow as a kind of national treasure and an important "lady to see" for anyone looking to enter the country's political fray.

After learning this much, I replied to M. Heartsworth's email and told him I'd love to meet him, and soon.

What I learned about him later was no less impressive. Heartsworth had studied environmental management in the UK before coming home to work for ZCCM. By 1990, he had become an environmental officer for the company, a job which made him responsible for managing the lead problem in Kabwe, or trying to. Dutifully, M. Heartsworth collected reams of data regarding the lead content in local soil and water supplies. He documented the lead levels found in the blood of local children and became increasingly distraught when ZCCM refused to act. The best they would do was distribute free milk to afflicted kids. The calcium found in milk can slow a body's absorption of lead, but solutions like this were akin to putting a Band-Aid on a chest wound.

True to his father's spirit, Heartsworth began a campaign to bombard his superiors with the damning data he'd gathered. The company ignored him at first. Then it took his driver away, making it difficult for him to visit Kabwe or any other site ZCCM had polluted. Heartsworth did not despair and continued to send his reports.

At last, one morning in 1996, a police officer showed up at Heartsworth's home and said, "Congratulations! You are going to a university in Canada to earn a postgraduate degree."

Heartsworth said, "Oh yes? I had no idea."

"Indeed!" the policeman said, grinning. "Your company has decided to do this for you. How nice of them. So. Here is your ticket. Your plane, as you see, will depart in three hours."

Heartsworth stared. "I can't leave that quickly," he said.

The policeman stopped smiling and said, "You will go."

As Heartsworth later explained to me, Zambian citizens know better than to argue with an armed policeman charged with dispatching official errands. Nervous as hell, he packed quickly and let himself be driven to the airport, where he boarded the waiting plane that took him out of the country. He would be gone for years.

"I will send you documents," Heartsworth promised. "You hold them. Keep them safe."

When the trove arrived, I made backup copies before sitting down to read them. With each page I turned, my stomach dropped. Here were several reports written for ZCCM in the mid-1990s by Clyde Hertzman, a Canadian doctor and epidemiologist. Dr. Hertzman noted that young children he had examined in parts of Kabwe had blood lead levels of seventy-five micrograms per deciliter or more, "the highest I have ever seen in a community sample." Some test results, he wrote, showed lead levels over 300, well into the imminently fatal range.

"The effects of lead [poisoning] mimic the effects of neglect and lack of stimulation," Hertzman wrote. "Intellectual function is reduced, and various 'neurobehavioral' barriers to learning such as irritability and distractibility increase. Literature reviews suggest that intelligence quotients decline by five to eight points for every ten microgram/deciliter increment in blood lead in early childhood." By Dr. Hertzman's final analysis, the blood lead levels he found in Kabwe "would translate into an average IQ reduction of ten to fifteen points [per citizen]."

It was enough to begin. I asked Heartsworth if he wanted to head up efforts to address the situation in Kabwe, and he agreed. He formed an NGO that I paid for called the Kabwe Environmental Restoration Foundation, and together we started plans to make a difference.

Next, I got in touch with my World Bank friend, Yves Prevost, who jumped into action with the World Bank's government contacts in Zambia. At the time, the World Bank was negotiating a large loan for work in another part of the Copperbelt.

"You have to include Kabwe in this loan," Yves told them.

The Zambian officials refused at first. Kabwe, they said,

was off limits. If you insist on including Kabwe, we'll drop the project.

"Fine," said Yves. "The project is off."

It was brinksmanship at its finest. Yves knew the state of affairs in Zambia as well as anybody; he knew they really needed the loan, and he was forcing government officials to face the realities in Kabwe.

After a week, the Zambians came back to the table and asked to renew talks.

"Fine," said Yves. "But first you need to watch this movie."

He took them into a conference room, locked the door, and made the Zambian government officials watch *Erin Brockovich*, starring Julia Roberts, from start to finish. He watched them squirm as the parallels became obvious.

"Okay, here's the deal," Yves said when the film was done. "Kabwe is no longer a secret. Pure Earth and the Kabwe Environmental Restoration Fund are out there, telling the world. You can't hide it any longer. You have to face it. Fix it. I'll find the money to help from our grant programs. Not loan funds—let me be clear. These are grant funds we can allocate, since Kabwe represents such a clear case of human emergency. We'll work together to look after the children and families in this blighted town. Deal?"

The Zambian government officials had no other option. Reluctantly, they agreed.

Once everything was settled, the cleanup project for Kabwe had plenty of funding—almost $10 million. With the World Bank in charge, I let Pure Earth take a back seat. We continued funding KERF until Heartsworth went off to pursue other projects. We also kept an ear to the ground through other NGOs we had met. But I basically left the project in the hands of the World Bank. In retrospect, I think this was a bit premature.

Ten years later, the World Bank project was done, the program closed. I wanted to see what had been accomplished, so I sent a team back to Kabwe and had them test the local children and the soil at 400 locations around the old smelter, including houses and playgrounds. The results we got were disturbing; Kabwe was still hopelessly toxic.

The World Bank money had paid for a lot of work. For instance, the toxic canal had been dredged, and lots of the worst contamination had been buried in landfills, which is the proper way to handle lead-contaminated soil. A great number of children had been treated for lead poisoning. Trees had been planted all over the community. But most of the money had been spent on designing the next steps, and no one had been there to push the full remediation work through the system. Designs for a full-scale cleanup are there, and ready to roll. They've simply been mothballed.

So we've taken up this cause again. This time, we want to initiate a full-scale project that will incorporate lessons learned in past endeavors. This time, we want to do work that will really look after the children of Kabwe. The World Bank is keen that we do this (the project has been on a slow backburner there), so now there is a bit of energy to go and finish what they have started.

Sadly, Kabwe is not the only site contaminated by lead; after my trip to Zambia, I began to find them everywhere. I also learned something startling about the way lead toxicity frequently spreads. The biggest culprit, apart from massive legacy smelters like the one in Kabwe, is car batteries.

Consider how ubiquitous car batteries are: one unit for every automobile in the world, of which there are more than a billion and growing, according to a 2011 report released by auto industry bellwether Ward's Auto. These batteries help start gasoline engines all over the world, and each unit contains sulfuric acid, plus about seven pounds of lead, which of course means that each unit is potentially toxic. They're safe when they're in our cars because the lead and sulfuric acid are contained by a plastic casing. In the wrong hands, however, they can prove disastrous.

Car services in the U.S. and Europe are required to accept old batteries whenever you buy a new one. The services send the old batteries to large, multimillion-dollar recycling facilities where highly trained workers carefully open the casings, extract the acid and lead, and recycle all components to make new raw ingredients. These Western facilities do the job right. When handled properly, a car battery can be almost completely recycled. Not so in the global South, where countries don't enjoy such technological luxuries.

In the world's poorer countries, less than half of all batteries are recycled safely. More often than not, backyard operators break open each battery unit with an axe, dump the sulfuric acid (full of lead) onto the ground, and take out the lead plates by hand. Next, they melt the plates in a big pot over an open fire. During this stage, they often burn the batteries' plastic casings for fuel. Imagine the nasty billowing black smoke produced from this kind of setup—but that isn't the worst of it. The stuff bubbling in the pot is partly metallic lead (which they pour into ingots and sell) and partly lead oxides.

Most backyard recyclers throw the lead oxide away. Since they often perform their work on open land, and often near rivers, the lead oxide usually gets heaped on the riverbanks. Mind you, lead oxide is the most toxic form of lead. It dusts up easily and blows throughout the community, infiltrating water and soil. In this manner, the backyard battery recycler ends up poisoning thousands of his neighbors, killing children in the process.

I call lead a scourge and make no mistake, I use this word intentionally. The definition of scourge—something that causes great evil, distress, or suffering—fits perfectly. Pollution from car batteries improperly recycled in backyard operations affects over twenty million people at hundreds of sites in dozens of cities around the world. Throw in the people affected by polluted legacy sites like the Kabwe smelter and the number of people lead affects globally rises to twenty-five million. That makes lead the single most virulent agent for toxic exposure around the world.

Let me share some stories with you.

In this book's introduction, I mentioned Seynabou Mbengue, who lost five of her children to lead poisoning. She had made a cottage industry of recycling car batteries in Dakar, Senegal, while her kids were underfoot. Each of them died from symptoms brought on by acute exposure to lead. Seynabou had no idea that lead or any of the materials she was handling were toxic. She is one of thousands who have lost children in this manner in recent decades.

Pure Earth became aware of the problem in Dakar through a friend at the World Health Organization. In a somewhat random turn of events, this friend was tipped off by a European doctor doing volunteer work in sub-Saharan Africa. The doctor had noticed that kids in one neighborhood were dying from a disease that the locals assumed was malaria. But the symptoms he noted didn't quite fit with malaria's profile. Fortunately, he'd had some training in heavy metal toxicity and decided to dig a bit deeper.

He tested a sampling of local children for lead poisoning and bingo, there it was. The levels of lead toxicity he found in their blood were off-the-charts high. He contacted WHO right away, and the agency quickly contacted us and asked that we act immediately. So we sent a team to Dakar armed with equipment to measure soil contamination.

We needed to work out where the lead was coming from. Logic told us that the contamination source must be somewhere in the community. It also, we surmised, had to be large enough (read: potent enough) to affect such a large population. The women who had lost children knew the cause, and led us to it. They escorted us to a gray sloping mound that rose about twelve feet off the turf in a patch of woods near their homes. Picture a hillock of gray dirt—which was lead oxide, a monument to battery waste—with occasional plastic battery plates and cracked plastic casings poking out of it here and there. This pile of waste had been dumped from earlier battery recycling efforts in an industrial area of the Dakar neighborhood, an incongruous and lethally toxic blight that the women had been digging up.

My team asked the locals how the hillock had come to be there. They reported that some nameless representative of an even more nameless industry had dumped the refuse years ago. They had been recycling batteries. Then, recently, emissaries from a Chinese company had descended upon the neighborhood. They'd heard about the pile and were offering what amounted to a handsome payment (by local standards) for anyone harvesting decent-sized chunks of lead that they would export.

Some women from the nearby communities took them up on the offer. What choice did they have? That part of the world was poor, and the Chinese were offering decent wages. They began making trips to the pile day after day, walking barefoot, as was the local custom, and wearing no protective gear at all as they sifted through the waste. In doing this, they exposed themselves to the toxins, of course, but they were adults; their bodies could weather exposure more capably than the children tied to their backs.

In this part of the world, nearly every woman of a certain age has a young child, if not several. The women tie their children in cloth bundles slung across their chests or backs and carry them everywhere as they go about their daily chores. In other words, these children—in most cases, infants—were very much in attendance as their mothers and caretakers worked with the lead oxide. So of course they were contaminated by proximity, killed slowly over a matter of months. Each trip to the contaminated hillock exposed them to greater and greater degrees of toxins, until their tiny bodies simply couldn't take it anymore and shut down.

By the time we arrived, about thirty small children had died this way. The mothers grieved their loss, of course, but death, and specifically infant mortality, is much more prevalent in the developing world than most Westerners can probably imagine. Besides, as I mentioned, the entire community blamed malaria for the deaths, which is also understandable since, to the untutored eye, malaria's symptoms closely align with those of lead toxicity.

Unaware of the true reason behind their children's deaths, the women continued to visit the mound. They even formed a co-op in order to maximize their labor. But the cycle had to be broken lest more children die. We began our intervention at once.

Our first order of business was to cordon off the area, stopping any further exposure to the lead oxide. We then launched an education campaign that quickly taught the locals about the dangers of this waste. The neighborhood began to support our efforts once its residents knew what was in the pile that people were mining, and what it was doing to their health and the health of their families.

In this case, as in so many cases Pure Earth addresses, human nature was on our side. There isn't a mother on the planet who wants to hear that she's responsible for making her baby sick, however inadvertently. Once the villagers were properly informed, keeping people away from the contaminated area became relatively simple.

Next, we began to raise funds, which we secured initially from our own resources and then soon thereafter from a wonderful NGO called Green Cross Switzerland. While this was going on, we sent emissaries, including myself, to the local offices of Senegal's Ministry of the Environment. Thankfully, the bureaucrats ensconced there saw value in what we were asking for and kept red tape to a minimum, which isn't always the case. With funding and proper permissions in hand, we then hired contractors, who brought heavy machinery to the site, dug up the contaminated soil, and carted it off.

The government granted us permission to use an abandoned mine located about ten kilometers from the affected villages as our depository. We packed the mine with the toxic material and sealed its entrance with a concrete plug so that no one could somehow stumble into a potentially nasty situation. Finally, we returned to the site where the lead oxide had been heaped and brought in many truckloads of clean soil, which we spread so the area could re-seed itself and eventually re-integrate with the surrounding countryside.

All this happened in a matter of weeks—rather quick when compared to other interventions we've executed in other parts of the world. But, of course, that was only Phase One of our plan.

During Phase Two, we identified danger spots throughout the communities proximate to the lead oxide dump—anywhere that dust from the pile might have blown into houses, backyards, schools, and community playgrounds. Pure Earth personnel arrived at each prospective location with a handheld X-ray fluorescence spectrometer. A device that looks like a ray gun from a 1950s science fiction movie, the spectrometer can identify and quantify the presence of certain elements, including heavy metals. With the XRF's help, we found hundreds of locations that required immediate cleanup and noted each one on a map of the region. We then began scheduling abatements: cleanups performed by hand to remove any lingering toxins that would render the area unsafe for habitation.

A major objective of Phase Two was to reconstruct a local economy for the women who'd once made their living scavenging lead from the pile. So we trained and paid those women to perform the cleanup work. They went throughout the villages properly garbed and masked, sweeping and mopping, wiping down every surface they could find three times and properly disposing of their spent cleaning supplies.

Phase Two took about six months to complete and I must confess, it was satisfying to turn the tables; the same demographic that had been afflicted by lead was now leading the cleanup effort. But what would happen to these people long-term? Once the cleanup was finished and all the lead removed, how would they make their living?

We felt quite strongly that all good efforts would come to naught if we didn't do something to address this problem. So we polled the members of the same women's cooperative that had mined the lead, asking them what other work they'd like to perform instead. What other skills do you have? we asked. Or what skills would you like to learn so the cooperative can keep functioning in a healthier trade? Resoundingly, the women said they could sew, so that's the angle we pursued.

It didn't take much. We purchased some sewing machines and set up the cooperative in a new facility, then partnered with local agencies who helped us land the first round of contracts. Within short order, the cooperative was off and running on a new, entrepreneurial path. Now, instead of poisoning themselves and their children by mining lead, the women had gainful employment at commensurate levels of income, sewing clothes to be sold at local markets.

It is not all good news, however. By the time this intervention was finished, hundreds of children had been acutely poisoned by lead. They will suffer some brain damage for the rest of their lives and nothing can be done about that—I only wish it weren't so. But we stopped the source of that horror in its tracks, and all for a total cost of something like $150,000. If all goes well, it will never happen again, though one must exercise constant vigilance. Here's an example of why.

About a year later, we received word that the government of Dakar had been approached by the mayor of the community where the abandoned mine resided. The mayor was furious that contaminated soil had been dumped in his district. It was dangerous, he said, which is why he had personally hired men to break the concrete seal, cart all the toxic materials off, and dump them on the side of a public road.

I had a hard time believing this when I heard it. At that point in my life, I thought I'd seen political grandstanding at its finest. Evidently not. Regardless, we had to correct the situation somehow, and fast. So we raised some more money, flew back to Dakar, and found the materials exactly where this wacky mayor had put them. Immediately, we cordoned off the area and covered the toxic materials with plastic tarps to keep them from blowing around while we formed a huddle and worked out our options.

In the end, we got the government's approval to dispose of the waste with what's called "local encapsulation." While this type of solution is quite common in the United States, it had never been done in Africa; we were the first to break this kind of ground, both figuratively and literally.

To create a local encapsulation, you dig a grave for toxic materials in the middle of an existing landfill dump. Once we were granted a host site, we brought in heavy equipment, sent it straight to the center of the dump, and had the machines dig a hole fifty feet straight down through the trash. We lined the hole with a special leak-proof sheeting, poured all the toxic materials inside, and capped the hole with another plastic liner whose edges we sealed around the first to create an inviolate sack of poison. As a final step, we replaced the garbage we'd dug out to make the hole in the first place and voilà, the problem was solved.

The total cost of this second intervention came to about $100,000—not a whole lot in the grand scheme of things, and worth every penny to any unfortunate person who might have stumbled over that soil after Mayor Witless upended it by the side of a well-traveled thoroughfare.

Lead contamination is hardly confined to Senegal or even Africa. As I've mentioned, we find it all over the world. Remember back in this book's introduction when I cited the polluted soccer field near Jakarta, Indonesia? We intervened there back in 2011. The details of that case were slightly different than the one I've just described, but the essence matches with awful precision: here was another site where dumping from backyard battery smelters had poisoned an entire community. We found Indonesia to be particularly rife with such places. Imagine our shock when the regional analysis we performed turned up no less than 250 sites polluted by battery waste in the Jakarta suburbs alone.

The township of Cinangka, where the soccer field that I mentioned resides, sits on a mountaintop in West-Central Java just over an hour's drive from Jakarta's raucous streets. Despite its cheerful inhabitants, brightly colored houses, and blissful weather, Cinangka is one of the most polluted neighborhoods in Indonesia. Between 1983 and 2006, illegal lead smelting formed the backbone of Cinangka's economy. During that interval, no less than thirty-two illegal smelters choked the township with black toxic clouds of lead. The atmosphere became so lethal that guavas growing from local trees turned the color of chocolate. Birth defects became common in the local populace, with many children showing signs of mild to severe developmental issues.

We were first guided to Cinangka by Alfred Sitorous, a young Indonesian who works for Pure Earth's lead abatement subcontractor in Indonesia. Alfred travels throughout Jakarta's far-flung city limits, visiting areas affected by lead contamination. In Cinangka, Alfred always stopped to talk with local children playing in the lead-contaminated soccer field. He would point to their feet and chide them—with good reason. Kids are kids, the world over. The ones in Cinangka loved running around barefoot. They couldn't fathom the danger they were in, but Alfred knew the truth.

The small white/gray squares he often found poking up from the soccer field's turf were half-buried battery separators. Lead-acid batteries make use of materials like rubber, cellulose, glass fiber mat, and polyethylene plastic to insulate positive and negative electrodes. Cut into squares, these separators get sandwiched between the lead plates to keep them from touching and shorting out. Over the course of a battery's life, the separators become saturated with sulfuric acid, lead residue, and other chemical by-products. They can be just as toxic to human beings and the environment as the lead components themselves.

Seen from Alfred's point of view, the soccer field was nothing but a crude landfill for millions and millions of separators. Which is why his XRF gun registered portions of the soccer field at more than 100,000 parts per million of lead. Compare this number to WHO's warning that soil containing more than 400 parts per million of lead can be hazardous to human health.

It was remarkable to watch Alfred work with the kids. He's practically a big kid himself. He'd drop to one knee and make barnyard noises—cows and chickens, pigs and cats—to call local children into a big circle and make them giggle. Then he'd lead them in songs and little dances before playing a seemingly impromptu game of Who Can Put Shoes on the Fastest? Squealing with delight, several barefoot children would scamper off to retrieve sandals they'd kicked into ankle-high grass. Dropping to their rumps, they would scramble to pull on their shoes while Alfred started a Name Game with the remaining kids.

Why the Name Game? Alfred was clever. By learning each child's name, Alfred was able to create a mental roster of kids who weren't wearing shoes. He'd take this list back to local officials who knew the soccer field's status and were working with us to address it. The officials would speak to the children's parents and remind them that, in Cinangka at least, children had to wear shoes if they wanted to stay healthy. To stay alive.

Our first measurements of the kids' blood levels found that they were already saturated with dangerously high levels of lead. It would have been best, of course, if they had stayed away from the soccer field entirely, but telling an Indonesian child not to play soccer is like telling a housecat not to slink. It's asking the impossible. Wearing shoes and washing hands frequently was the best that we could hope for until a proper solution was found.

So how did we intervene? It took a lot of patience and a very big pair of scissors to cut through Jakarta's web of red tape. In the end, we dug up the contaminated soil and shipped it off for interment in another local encapsulation we created. Total cost? About $250,000. Again, not very expensive—certainly not when you consider the project's total impact and all the lives it ended up saving. But there was another unexpected and satisfying result to this project.

Our intervention worked so well that it opened the eyes of Jakarta's government to the levels of contamination festering within its suburban limits. Which isn't actually true, of course—the government knew all about the levels of contamination. Perhaps more plainly, we could say our intervention showed them that a) other people were paying attention now, and b) these people were willing to help them address their pollution problems, provided they stepped up to the plate when called. In other words, we were offering help, so why not take it for everyone's good?

Here's a final lead project I'd like to tell you about, which took place in Vietnam. At some point, during its long history of Communist rule, the Vietnamese government designated a system of craft villages. The name will strike some Westerners as ironic. In American minds, for instance, the term "craft village" might conjure images of artisans whittling away, making knickknacks out of wood, engaging themselves in handicrafts, furniture-making, what have you. Not so in the Vietnam of twenty years ago. Back then, being part of a craft village meant that the government had allocated your whole town and all of its personnel and resources to one particular industrial process. Certain villages recycled tin cans, others just did plastics, and so on. If you grew up under an unlucky star, the industry your village handled might be very dangerous, indeed.

We were contacted by the Vietnamese government, which we found very collaborative and very cooperative, and with good reason. They were keen to reduce levels of lead toxicity in the Vietnamese people. As luck would have it, the government had devoted an entire village to recycling car batteries. We went to this place (which will remain nameless), aided by funding from the European Union and the World Bank, through the Global Alliance for Health and Pollution. Sadly, we weren't surprised to find a situation even worse than the one we'd registered at Dakar.

Lead oxide could be found throughout the town. The main street was full of lead waste. No wonder the people's blood levels were high. The contamination lay everywhere. It was literally underfoot as they passed from home to market to school to work and play, and back again.

This time, we ended up using a different abatement method, a technology called capping that has also gained the approval of the U.S. EPA. First, we measured soil lead levels throughout the town. Some were off-the-charts high, while others were simply very high. At any rate, the amount of soil that we needed to remove was enormous. We could only take the worst of it away; we buried it as we had in other projects. But the remaining materials, with lower levels of lead, we left in place and covered with first a plastic barrier, then with more soil.

The barrier and soil cap served two functions. First, it kept the contaminated soil from spreading. Second, it would warn any future generations who encountered the cap inadvertently. A plastic sheet buried under the ground? In essence, it would be saying: *Beware! This was put here for a purpose, and breaking the seal serves no one's best interests.*

Once the cap was in place, we covered it with a couple of feet of fresh soil and paved over that with concrete. Consider the beauty of this technique. The landfill wasn't hidden anywhere. It was right out in the open in a place where it likely would never have to be moved again: under a road. It also served a vital purpose. The landfill had become something people used every day. This kind of capping also offered the advantage of being less expensive than most other intervention types.

Scientists from the University of Washington analyzed the blood chemistry of children from the village and found a pre-intervention average of fifty-five micrograms of lead per deciliter. They took more samples three months after the cap was installed and noted that this average had dropped to about twenty micrograms per deciliter. The safe limit as defined by most health organization is five micrograms, so clearly we still have a ways to go. We were heartened, however, to see chemistry levels moving in the right direction so rapidly and to such an extreme.

Senegal, Indonesia, and Vietnam. Three projects that successfully and significantly reduced lead exposure to local populations without spending ridiculous amounts of money. There are lots of other projects like this, where a small amount of money has done a lot of good. Those are the kinds of results we shoot for at Pure Earth, and I'd like to point out that they all shared common denominators. First, the communities we visited were aware that they lived in a highly toxic environment, but they had no idea how to deal with their problem—what actions they should take, what technologies they should embrace, how much it would cost, and where they could go to get help. Second, the solutions we eventually chose to implement weren't glitzy, but rock solid, practical interventions. Third and finally, in each case, we spawned lots of copycats of our techniques, a phenomenon that pleases us immensely. When other groups, agencies, and governments started to mimic the moves we made on local and international scales, we knew we were doing something right, if only by raising awareness that brown problems could be handled effectively and without great expense.

I'll spend a lot of time furthering this idea of raising awareness in the last part of this book. For now, however, keep reading. We've got plenty more pollution types to go.

This journey is just getting started.

CHAPTER SIX

Cold War Killers

After my first trip to Horlivka (the one I described in this book's introduction), our local coordinator, Vladimir, and I returned to Kiev. I was so upset by what I had seen that I knocked on the door of the European Commission, which was already funding us with a decent grant of about a million euros or so.

"Look," I told my contacts there, "I've just been to Horlivka, and I've seen the chemical plant. We need to do something about this right away."

Bureaucracy stepped into standard operating procedure and I found myself on the receiving end of a whole bunch of throat-clearing, mmmmm-hmmmmms, and variations of "we don't have the money." Which, of course, I had kind of expected.

"Right," I said. "Tell you what. Let me put together a summary of the conditions I saw at the chemical plant. I'll send it to you. You read it. Then perhaps you can tell me how we can let this situation stand."

While doing that, I scraped together about $50,000 from our board of directors and sent Vladimir out to organize a cleanup crew.

"Fifty thousand American dollars," Vladimir said. "That's not much to work with."

"I know," I said. "But it sends a message. I want these agencies to see that we're not just sitting around on our hands. We're going to do this with or without them."

Vladimir and Sergei, our other Ukraine coordinator, connected with the local government agencies and began doing some serious legwork. They asked the local folks, "How far can we stretch this money? In order to do something productive—to reduce the risk the site currently poses—it needs serious work. But where do we start?"

Very quickly, Sergei and Vladimir came back to me with a plan. "We could do some repackaging work," Vladimir said. "The MNCB is lying around all over the place. We could gather up the worst of it and seal it in nice tight drums to keep it from leaking again."

Vladimir used the word "again" because, while researching the Horlivka plant further, we had discovered an earlier incident that realized our worst fears. After years of leaking from decrepit tanks and drums, MNCB had slid down some cracks in the earth and infiltrated a series of mineshafts running near and under the site. Three miners had been working down there; all three were killed instantly. Realizing that the MNCB had volatized and vaporized, local authorities quickly sealed the shaft. That whole area, we were later told, was a tomb of toxic gas from which the miners' bodies were never recovered.

"Picking up the worst of the MNCB is a good start," I said. "What barriers do we face?"

Vladimir made a face. "There are many," he said. "Regulations. Permits. Sanctions. But I think I know someone who can help."

A few weeks later, I flew back to Ukraine and met with Alexandr Brishnalov, a senior director with the Ukrainian Ministry of Environment. He was an old-school Soviet bureaucrat who talked big, acted big, and was probably (my instincts told me) ever so slightly corrupt.

Soon after my first meeting with Brishnalov, he took Vladimir and me out to dinner. It was one of those fancy, touristy restaurants that cater to Russian businessmen, since basically they're the only ones who can afford them. It was the kind of place where a single plate cost $100, in a country where you could feed a whole school for less. A place designed to make an impression—though I have to admit, the food was great.

The whole time we were in there, Brishnalov kept his driver waiting out front with the engine idling. He was making a calculated show of being the Big Man in Charge, the guy who was Taking the American Out on the Town. Looking back, it was sort of ridiculous, as was the fact that, no matter what oddities Brishnalov perpetrated, I found myself actually liking him. As big as he talked, he came up with solutions to many of the dilemmas we faced, which of course we appreciated.

When the bill finally came, Brishnalov took the check and paid it—in cash, I noticed. I remember wondering why he had done that. We had come to him for help; by all rights, I should've picked up the check, which I was more than willing to do. As a kind of conciliatory gesture, I offered to buy some whisky so we could toast the eventual cleanup of Horlivka. We each had two shots and I called for the bill. Imagine my surprise when it came to more than $250.

I called the maître d' over to our table. "I think there's been some kind of mistake," I said.

The maître d' checked the bill and shook his head. "No mistake," he said. "How would you like to pay?"

"I wouldn't," I said. "Two hundred and fifty dollars for two rounds of whisky is outrageous."

At which point four big thugs materialized from the shadows. They stood over us, glowering. The message was clear. I paid the check and we left the place. Overall, it was a very Russian experience.

So we had some meager project funds, plus assurances from

Brishnalov that we could work at the factory campus. This last point was especially important, since the site was being guarded by Ukraine's equivalent of the National Guard. Despite the conditions the site was in, the place was still, technically, a secret government facility.

These guards carried Kalashnikovs and had very bad attitudes. As we soon discovered, they took their job quite seriously, which—come to think of it—was all well and good. Access to such a dangerous place *should* have been restricted.

But this, of course, was our project site. We had to be granted admittance if we were going to remove the MNCB. The guards made it seem like we were the ones who put the stuff there in the first place when, in reality, we were the ones who had come to remove it. And, of course, we had come up with the funding.

At about this point, I got another idea. Why not approach big companies like Dow Chemical and ask them to help with the project? As luck would have it, a member of my board of directors had introduced me to some of the company's top brass. Through their trade group, Dow gave us about $100,000, which we were grateful for, of course. But then they went a step further and loaned us some of their top chemical engineers. The scientists accompanied us to the site, surveyed it with us, and offered their expert opinions on how best to clean up the place.

These same experts were also one of the first groups to apprise us of the fact that the plant may once have been used for seriously bad purposes. This subject didn't come up until the day we entered the site to begin our preliminary cleanup. One of the scientists took Sergei aside and spoke to him in a voice pitched low, so as not to be overheard.

"I wouldn't confront your Ukrainian colleagues about this now," he said. "But with this much MNCB lying around? And all the other chemicals present at this site? You might want to seriously consider that this place was a chemical weapons facility."

Top: **The author on a riverboat on the Purus River, 1989.**

Bottom: **An aerial view of the Amazon rain forest deforestation, 1989.**

The author with the Great Forest team (*top*); at a waste dump in Phnom Penh, Cambodia (*middle*); and with Peter (Oz) Hosking in the Great Forest offices (*bottom*).

***Top*: The author with local children in Angkor Wat, Cambodia.**

***Bottom*: Children in a contaminated creek in Thailand.**

Waste dump scavengers wearing improvised protective gear in Phnom Penh, Cambodia.

Top: A neon-blue effluent stream from nearby factories winds its way through a residential neighborhood in Dar es Salaam, Tanzania.

Bottom: Water pollution in the Marilao River Basin in the Philippines.

Top: In Agbogbloshie, Ghana, locals burn scrap wires and other electronic waste (e-waste) at informal processing areas to extract the valuable copper within. The burning poisons the landscape and releases large amounts of toxic fumes into the air, which settle into the water and ground and contaminate food sold in the nearby markets. Agbogbloshie is one of the largest e-waste dumpsites in Africa.

Bottom left: Recyclers dismantle e-waste in Agbogbloshie.

Bottom right: E-waste scavengers in Agbogbloshie.

Top left: Children play soccer barefoot on a lead-contaminated field near a primary school in Cinangka, Indonesia. The soil measured hundreds of times above the U.S. EPA standard.

Top right: The soccer field in Cinangka during Pure Earth's remediation.

Bottom: Dozens of scavengers roam mine tailings looking for lead in Kabwe, Zambia. Lead poisoning affects nearly the entire population of Kabwe—around 200,000 people.

In the former Soviet town of Horlivka, now in Ukraine, a dilapidated TNT factory sat abandoned for more than a decade with hundreds of drums of leaking toxic chemicals near tons of explosives. The plant was a toxic dump in the center of a city housing 260,000 people.

Top: **TNT crystals lie ominously inside pipes at the Horlivka explosives factory.**
Bottom: **Traces of TNT by-product litter the grounds of the abandoned factory.**

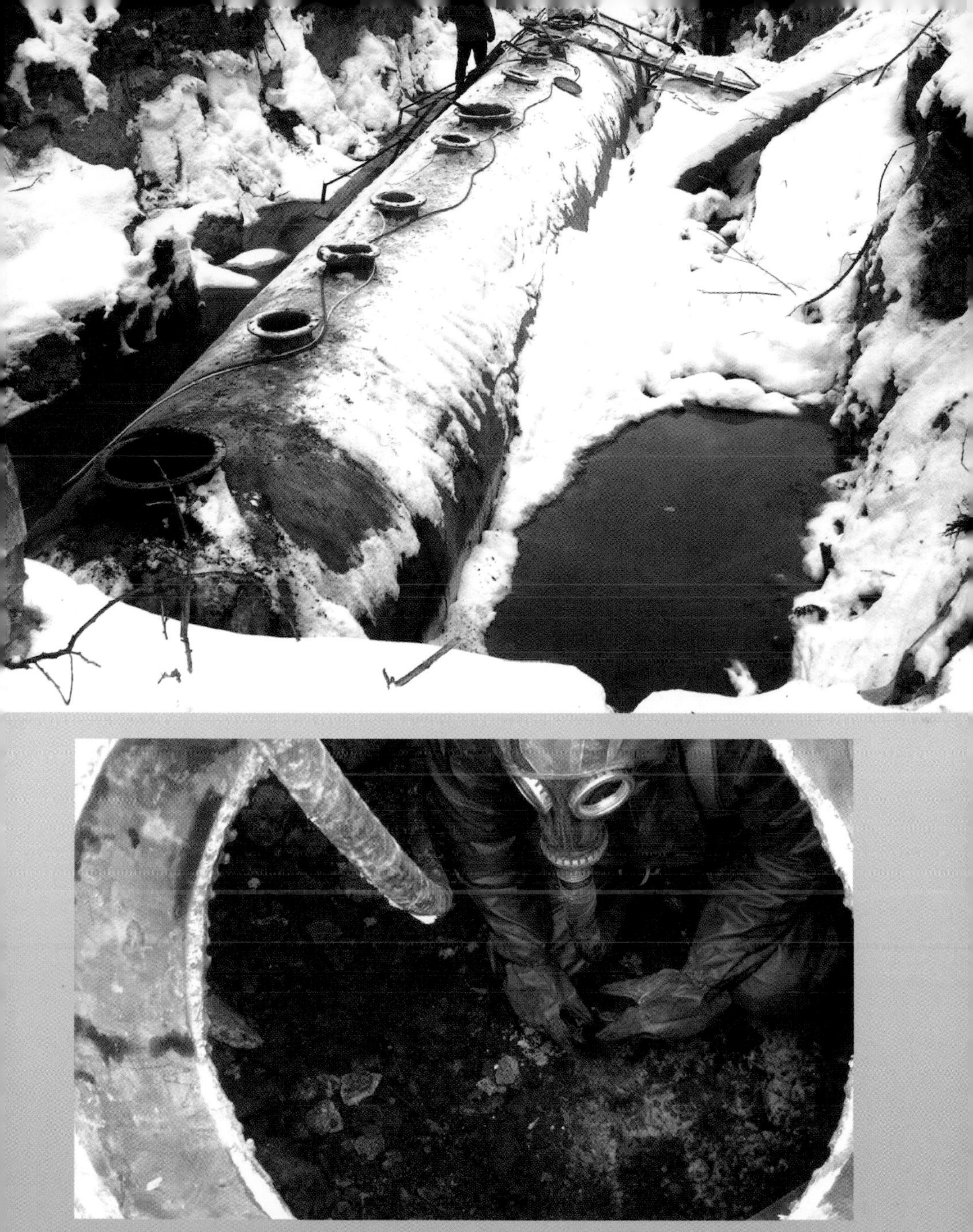

Top: These tanks were used to hold TNT during emergency flushes of the explosives production equipment at the Horlivka plant. When the factory was abandoned, tens of tons of TNT were left in the tanks. The red water around this tank indicates TNT that has leaked out into the soil and groundwater. TNT is not only a dangerous explosive, it is highly toxic as well.

Bottom: A worker in a hazmat suit works inside one of the storage tanks at the TNT factory. This was one of the most dangerous cleanups Pure Earth has ever conducted.

Top left: Toxic mercury is used to extract small bits of gold in the artisanal gold mining process. After processing tons of ore and burning off the mercury, these tiny bits of gold are the miners' prize. The excess mercury is often dumped on the ground or in nearby waterways.

Top right: A miner holds balls of amalgamated mercury and gold in his hand, demonstrating the most common way to recover gold in small-scale mining.

Bottom: The mercury amalgam is commonly burned in backyards, as shown here in Cambodia, in kitchens or in gold shops. Women and children are often close by, not realizing the danger posed by the invisible mercury vapor.

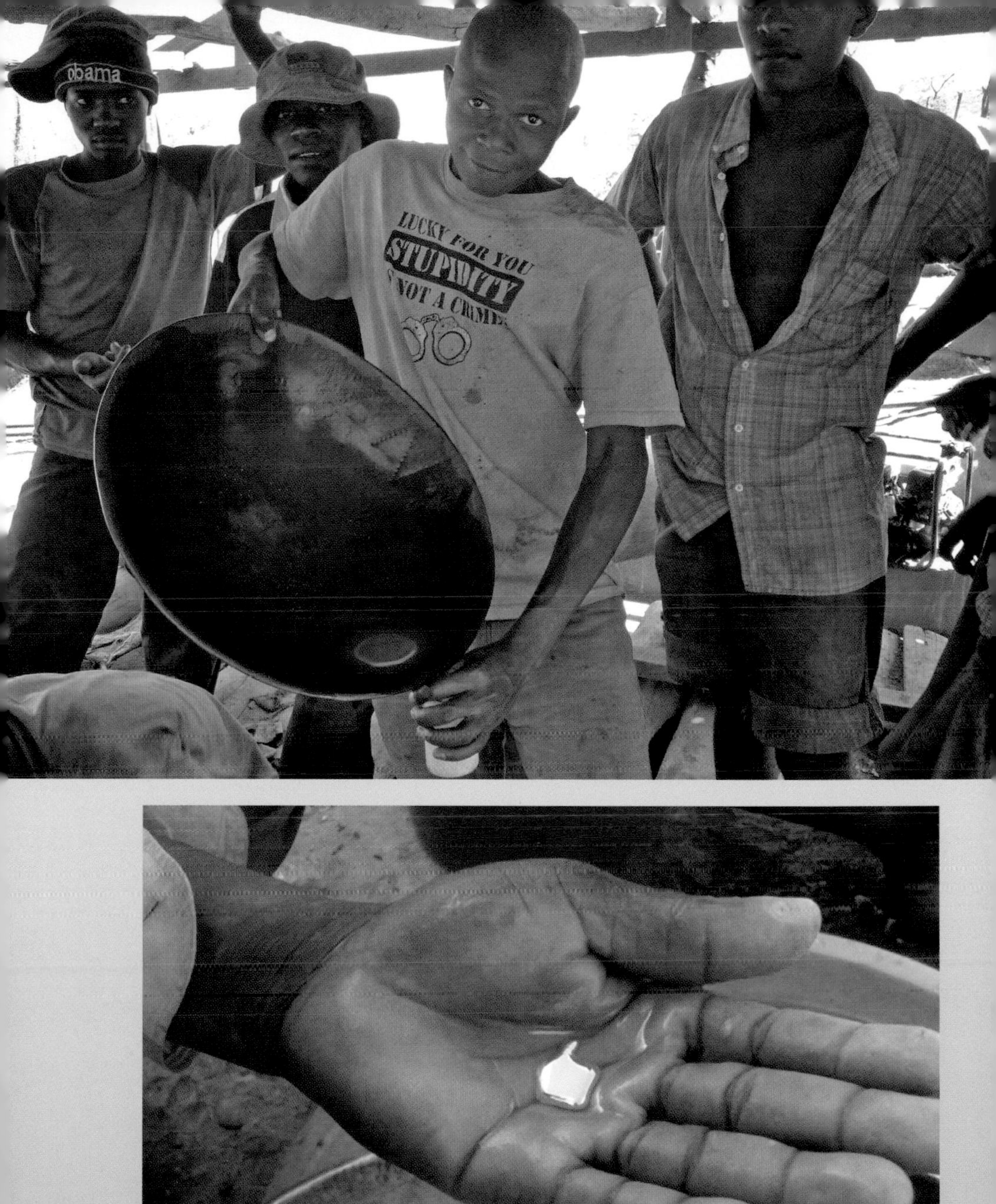

Top: African gold miners pour the amalgam of mercury and gold into a bottle. Next, the liquid will be heated to recover the gold, and the mercury boiled off into the environment.

Bottom: Toxic mercury is widely used to recover gold in small-scale mining. Artisanal gold mining is now the single largest source of mercury released into the environment by humans.

***Top*: A young gold miner in T'Boli, South Cotabato, in the Philippines.**

***Bottom*: Two young miners work on concentrating the ore to capture tiny grains of gold in Indonesia.**

Top: A pesticide burial site in Azerbaijan, where toxins are brought in from different parts of the region and encased in concrete "coffins." There are about 298 concrete containers here filled with DDT, chlordane, calcium cyanamide, and calcium arsenide. Large cracks have developed in the concrete containers, and the contents are leaking and polluting the nearby pastures.

Bottom: A dilapidated pesticide storage site in Somaliland, where toxic chemicals were leaking from old, corroded barrels.

Top: A child on top of an illegal waste dump in Nandesari, the "Golden Corridor" of industry in Gujarat, India, that hosts thousands of chemical, textile petrochemical, fertilizer, and pharmaceutical industries.

Bottom left: Children play in an informal drum recycling facility near Vadodara in Gujarat, where the fumes from burning waste are stifling.

Bottom right: Pure Earth advisor Dr. Jack Caravanos and Concept Biotech president Dr. Suneet Dabke sample the contaminated waste dumped into a river in Muthia, India. Approximately 60,000 tons of sludge from effluent treatment plants and other untreated waste have been dumped along the boundary between the industrial estate and Muthia village over the last decade. These hazardous wastes leach into the groundwater, and monsoon rains wash and spread the contaminated water over wide areas.

Top: **A young coal miner in Linfen, China, which has some of the worst air pollution in the country.**

Bottom: **The steel-producing mecca Magnitogorsk in Western Russia was responsible for half the Russian tanks created during WWII. The industry released 650,000 tons of industrial waste—including 68 toxic chemicals—and polluted some 4,000 square miles of Russia. The local hospital estimates that only one percent of all children in Magnitogorsk are in good health.**

The air in Norilsk, Russia, is so thick with pollution that no vegetation will grow within a thirty-mile radius of the city.

Top left: An abandoned childcare center in Pripyat, near Chernobyl—the site of the worst nuclear disaster in history, which emitted more than 400 times the amount of radioactive material as the Hiroshima bombing.

Top right: School children in Mailuu-Suu, Kyrgyzstan, are exposed to water contaminated with radioactive particles, a toxic legacy of the town's uranium mines. Pure Earth has installed water filters in the town's schools and hospitals to protect the most vulnerable.

Inset: Radionuclides in the soil at Muslymova, Russia. The reading is over 500 times what is considered a safe level.

Bottom: An abandoned amusement park in Pripyat, near the Chernobyl nuclear power plant.

***Top*: A small roadside tannery in Bangladesh, one of hundreds that operate using toxic chemicals with little to no regulations.**

***Bottom*: A Nepalese woman shows the results of using water contaminated with arsenic.**

Top: Polluted wastewater contaminates a public pathway in Ranipet in Tamil Nadu, India. Although the chemical factory responsible for the pollution has been shut down, its toxic legacy remains. The yellow water tests at dangerously high levels for hexavalent chromium.

Bottom left: A man in India holds up a glass of water drawn from a well that has been contaminated with chromium. Billions of urban and rural people depend on groundwater for their everyday needs. In many polluted places, groundwater is contaminated by the improper dumping of industrial waste.

Bottom right: A cup of chromium-contaminated water drawn from a well in India.

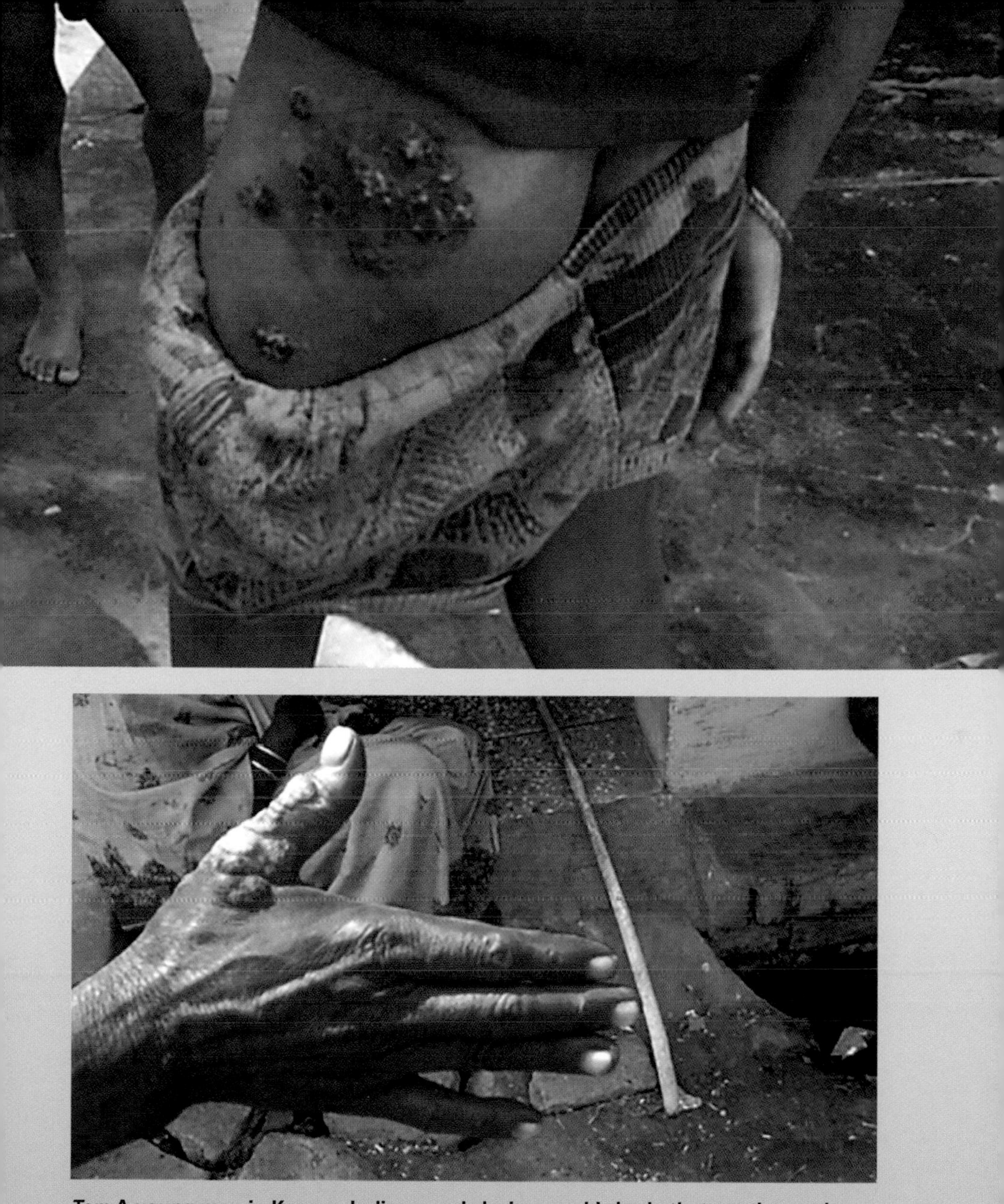

Top: A young man in Kanpur, India, reveals lesions on his body that are the result of exposure to hexavalent chromium. The lesions appeared after he attempted to water local market gardens with effluent from a nearby wastewater treatment plant.

Bottom: A woman shows the effects that contaminated water has had on her skin in Kanpur. The village sits right on top of a plume of hexavalent chromium emitted by toxic sludge from an old chemical plant that once supported tanneries. Flammable methane trapped inside the sludge catches fire during the hot summer months, releasing harmful toxins into the air; summer heat and winds also distribute dust particles from the sludge that are harmful when inhaled. Chromium from the sludge leaks into the river, subsoil, and groundwater—the primary source of drinking water for the surrounding community.

Top left: This contaminated well is the only source of drinking water in this village in the Sukinda mining area of Odisha, India.

Top right: Children gather contaminated water from their village's well.

Inset: Children play on top of chromium-contaminated tannery waste in Bangladesh.

Bottom: Garbage and tannery waste heaped on a public walkway in Hazaribagh in Dhaka, Bangladesh.

Sergei asked what kind of chemical weapons might have been made there.

The engineer made a point of not meeting his eyes. "Sarin gas," he said. "Probably a few others."

Quietly, Sergei began to make inquiries. He soon discovered that Horlivka's citizenry, the plant's former employees, and the national government of Ukraine did not want to talk about it. When pressed, they claimed that the MNCB found at the plant had been used as chemical feedstock in the manufacture of cleaning products and household items: paints and paint thinners, dyes, and the like. The mere mention of MNCB's use in chemical weapons brought an abrupt end to the conversation.

Vladimir and I asked Sergei what was going on.

Sergei had grown up in eastern Ukraine. He shrugged.

"We became an independent nation in 1991," he said. "But the Soviet influence in this area should never be underestimated, even now. Back then, we had to do what we were told to do. And now we must live with consequences."

◆

The first day we all set foot on the site, several of our volunteer scientists looked visibly shocked.

"I've never seen anything like this," one of them murmured.

Another agreed. "I've worked in lots of countries, seen a lot of really bad places. This has to be the most dangerous. Please tell everyone to be especially careful."

This was easier said than done. The floors of the buildings we were working in were strewn with crystalline TNT. The stuff was everywhere: caked in pipes, packed into corners. An act so simple as knocking over a pipe could easily cause a spark that would blow the whole place sky high.

We were appalled when we noticed our hired Ukrainian workmen drinking vodka and smoking cigarettes. Not that this was unusual behavior for your average Ukrainian construction site. Under our particular circumstances, however, it courted disaster.

Frankly, I thought it was a wonder that the place hadn't blown already. When our team first arrived, half the roof on the main TNT production building had collapsed. The damage was fairly recent, we noticed. Constructed of steel and concrete, the roof had buckled and fallen some sixty feet onto the machinery where the toluene was nitrated—in other words, where the TNT was created and made volatile. The tanks of these great machines still had plenty of TNT inside them. Venturing anywhere close to them was incredibly dangerous.

Regardless, we set to work packing up the worst of the MNCB we could find. In the end, we filled about fifty barrels with the stuff, which we capped nice and snug and readied for transport. Turns out, this gesture was enough to shake the trees. Returning to the agencies we'd previously approached, we presented our progress. Thankfully, things started to happen.

The Delegation of the European Union said they would climb on board and that they wanted to get the Swedes involved, too. Through the Swedish International Development Cooperation Agency, we were told that they'd be willing to help. Then our friends at the Swiss-based nonprofit, Green Cross Switzerland, offered to chip in.

Everyone said, "Yes, okay. We're in. But Ukraine has to put some skin in the game, as well."

At which point, we found ourselves once again working with Brishnalov, only this time in an official capacity. He was appointed by the Ukraine government as their local project overseer, which made perfect sense. After all, he was the attaché to the Ukrainian government, and therefore the person most qualified to shepherd the whole project through the maze of proper channels and compliance.

It was Brishnalov who pieced together an additional $5 million from Ukrainian resources to fund Phase One of Horlivka's cleanup. Phase One would essentially prolong and complete the preliminary MNCB cleanup our group had initiated. Under the Phase One guidelines, the barrels of MNCB we planned to fill would be taken to Poland for incineration, along with some others already present at the site. And here's where things turned especially weird.

When we mentioned that we needed a hauler who could truck the MNCB and destroy it safely, Brishnalov said he knew just the firm. He wrote a contract with an Israeli company that immediately tried to push Pure Earth out of the picture.

Basically, they wanted the money we would get in the project's second phase: removal of the TNT once the MNCB had been carted away. The head of the Israeli firm kept hounding us, saying, "We're the guys who should be doing this work, not you. You have no idea what you're doing; we do. Those funds belong to us."

Well. We checked these guys out and soon understood. Don't get me wrong. For the most part, they did their work just fine. Mysteriously, however, this one Israeli firm ended up winning all the environmental contracts in Ukraine. Hmmm.

We turned to our local contacts for the skinny. "What's the deal with this firm?" I asked.

"Totally corrupt," said Sergei. "Likely, they kick back money to all sorts of people. Including . . . "

I waited. "Yes?"

"You can guess," he said.

It took a lot of bobbing and weaving, but eventually we countered every insistence the Israeli firm made. The funds for Phase Two, we pointed out, were public monies sourced from the European Commission and other agencies. We couldn't just dole them out willy-nilly. We had to follow strict methods of compliance—UN purchasing regulations.

That one phrase, "UN compliance," seemed to knock the wind from their sails. We had to use it again and again, but it finally did the trick. Eventually, the head of the firm stopped asking about Phase Two funding. And one afternoon soon after that, his workers started the process of packaging and loading barrels of MNCB onto trucks and sending them off to a high-temperature incinerator in Poland.

A week or so later, I woke up to find an urgent message on my iPhone from Drew McCartor, our New York-based project manager. His email included a breaking news report. During the previous evening, a vehicle owned by the Israeli trucking firm had gone off the road and hit a tree near a small town about 100 miles west of Horlivka. The truck was reportedly carrying multiple tons of MNCB, enough to kill every living person in the town and a good deal beyond. Local authorities were still trying to assess the extent of any leakage, as well as who was responsible.

I called Drew immediately.

"Holy shit," I said.

"I'm on it," said Drew.

So too, as it turned out, was Brishnalov.

It was a tense morning, but we heard good news by the end of that day: our teams had packaged the MNCB so well that none of it had leaked, despite the crash. A second crew was able to load the barrels onto a new truck and continue to Poland. Authorities, however, were still investigating the precise nature of the accident. It turned out that the first truck's driver had been drunk while trying to negotiate the road.

"Not to worry," Brishnalov told us. "I will handle this."

And then he disappeared. Or perhaps it's more polite to say that he left the country without notice. We had no idea where he went, or why. One day he was there and the next, he was gone. According to the rumors we heard, he fled to cut his losses. Evidently, whatever kickback he'd received from the Israeli firm hadn't been enough for him to stomach their special brand of incompetence.

With Phase One finally complete, we commenced Phase Two: cleaning up the TNT. Despite our success cleaning up the MNCB, the second phase proved especially difficult.

With Brishnalov gone, we had no one to steer us through the Byzantine landscape of Ukrainian environmental regulations. There were permits to secure from several different agencies. Special dispensations to obtain. Apparently limitless fees to pay. A parade of government entities buried us in a blizzard of papers and complex contracts. At one point, the Ministry of Mines got involved for reasons that still seem odd to us. And all these procedures had to be completed before we could initiate our search for appropriate subcontractors.

Once we had lined up a labor force, our problems, if anything, magnified. A lot of the TNT at Horlivka was stored in two main tanks buried in front of the main nitration building. We had to pull those tanks out of the ground to get the TNT out of them. When we did this, groundwater immediately flooded into the craters. Actually, it's hard to call it groundwater, since the fluid was blood red. The craters looked like gaping wounds in the earth. TNT had leaked out of the tanks for years, polluting the countryside. It had permeated the groundwater tables from which the city drew its drinking water.

Our engineers saw this and shook their heads. Once we had cleaned up the MNCB and the TNT, they informed us, we should propose a Phase Three for the project: cleaning up Horlivka's drinking water. This is something that Pure Earth would still like to do but, as you'll see in a moment, the current conditions in Ukraine make the future of such an effort uncertain.

The workers we hired had to lower themselves into the holding tanks and literally scrape out the TNT with pressure washers. It was intensive, hands-on work, not to mention incredibly dangerous—entering a confined space filled with decayed explosives and agitating them. We insisted that the workers follow strict safety protocols designed by the engineers from Dow and other firms. The workers met our exhortations in typical Ukrainian fashion: with indifference. They followed the procedures we outlined, but did so begrudgingly.

Phases One and Two of our Horlivka intervention took a number of years in total. We got to the point where we had cleaned up all of the acutely dangerous stuff—enough material to take the place off the serious problem list. The MNCB was taken away and destroyed, while the TNT was snugly packaged and prepared to be incinerated or composted. The worst of the buildings had been dismantled and the project was in its wrap-up process. Soil and groundwater problems would need to be dealt with at some point along the line, but the Horlivka site no longer posed the threat of exploding and killing thousands of innocent people. We considered this a win.

Then, at some point in 2013, we noticed that all the guards had disappeared. We saw no more uniforms when we entered the site. No more Kalashnikov rifles. We felt as though something odd was afoot, but had no idea what it was.

Regardless of why the guards had left, we found it simply unacceptable to leave those stores of repackaged TNT open to the public. So we hired private security personnel to monitor the buildings in our absence.

Then came the Russian-led invasion of Ukraine in late February 2014. Since then, Pure Earth hasn't been active in the region. The place has become a war zone. Through sources on the ground, however, we know the following to be true as of this writing:

It appears that nobody is guarding the plant anymore. Our private security team simply stopped reporting for duty.

Some of the town officials we dealt with on a routine basis have been reported missing, probably kidnapped. Many more have fled the region. We know for certain that at least one was killed.

Although our project stabilized the Horlivka plant, it couldn't completely obviate the danger. Last we checked, some of the explosives that we removed were still on-site; we didn't have time to remove and incinerate them before the conflict erupted. At this point, we have no idea what's happened to them. They might have been stolen. They might still be there. They may have blown up. We have no idea whatsoever.

Being in the dark on this matter is especially frustrating since Horlivka was recently confirmed as a pro-Russian separatist stronghold. Both geographically and ideologically, it sits at the center of this conflict.

In mid-2014, the Ukrainian government tried to pacify Horlivka by shelling it with unguided ordinance. This is ill-advised, to say the least. Beyond the dangerous substances Pure Earth dealt with, the city hosts many landfills crammed with toxic materials. Explosions would agitate these materials and perhaps release them into the local environment.

Some of these sites are well-known, like the ammonia plant situated directly adjacent to the site we worked on. This plant has huge, aging tanks full of chemicals, though we believe the ammonia itself, the most dangerous of them, has been properly drained. More frightening to me, however, are the unknown sites—graveyards of chemicals, munitions, and explosives left over from the Cold War era, buried and forgotten, but no less deadly for all that. One misplaced explosion is all it would take to unleash a world of horror on an unsuspecting populace.

As dire as all of this probably sounds, Horlivka isn't the only environmental blight the Soviets left behind. The town of Leonidovka is another that warrants scrutiny.

Throughout the 1950s and '60s, chemical weapons of mass destruction were improperly dismantled and hastily buried in a forest ten kilometers northwest of a tiny rural village called Leonidovka, in Russia's Penza Oblast. Anyone walking through that copse of verdant birch and pine might scuff the soft loam floor with the soles of his boots and uncover the nose cones of World War II-era chemical bombs. Each piece of ordinance contained deadly substances such as yperite, a type of mustard gas, and lewisite, a colorless, odorless arsenic compound powerful enough to penetrate rubber and cause severe chemical burns, liver necrosis, and death.

As reported by the *Washington Post* in August 1998, the buried cache at Leonidovka contains sufficient quantities of WMD to wipe out every human being on the planet, assuming proper distribution.

That probably sounds bad enough. But a few years after the bombs were interred, the soil they lay in began to blacken and take on a metallic aroma. Nearby trees turned rotten and toppled. Plants grew skeletal, withered, and died. Eventually, even the grass disappeared, leaving a circular scar like a burn on the earth that slowly started to spread.

No one outside Leonidovka paid much attention, which is hardly surprising. In the greater scheme of Russian history, Penza Oblast has kept a quiet profile. Located about 300 miles southeast of Moscow (about midway between that city and the border of Kazakhstan), it features excellent farmland renowned for its output of beets and sunflowers, as well as pigs, cattle, chickens, and eggs.

But the locals knew what was going on. They had intimate experience with the bombs dating back to the end of World War II. At one time or another, the Russian Air Force had employed most of the village to produce, move, or store the WMDs. And the locals have not forgotten.

Years after the Cold War ended, Leonidovka's elderly residents still recalled the crude, inappropriate procedures they were given to perform their jobs. Year after year, they filed negligence claims which, year after year, fell on deaf ears. Doubtless, the Soviet military oligarchy understood what was at stake: Leonidovka was no isolated incident. It was the tip of a deadly iceberg, one site out of dozens and perhaps hundreds for which the Soviets were culpable.

Gradually, the truth began to reveal itself. In the early 1990s, during talks that led to the adoption of the UN's Chemical Weapons Convention, Russia disclosed its possession of some 40,000 metric tons of chemical WMDs stored in what it described as seven different locations. Leonidovka was one of them. About eighty percent of these weapons, Russian officials said, were fully weaponized lethal nerve agents, such as Sarin, VX, and Soman, fifteen million pounds of which were installed in aviation bombs and stored behind a walled military compound, also near Leonidovka. The remaining twenty percent of the arsenal was a hodgepodge of mustard gas, lewisite, and combinations thereof. This announcement stunned the world.

Activist Lev Fedorov told the *Washington Post* that, by his estimation, his government improperly disposed of some half a million tons of chemical weapons between the close of World War II and the late 1980s. Many were produced by Mother Russia and intended for use against Hitler's Third Reich. Other assets represented captured Nazi ordinance which, while different in design, posed no less of a threat to humanity.

Soviet officials had ordered portions of this combined horde sunk in the Baltic Sea and Sea of Japan, while others were hidden in unmarked, undisclosed, and no doubt improperly prepared graves across the former Soviet state. In other words, over the years, the Kremlin has buried so many chemical weapons that it has lost track of how many exist, and precisely where they are located. This situation, in turn, has touched off one of the foremost instances of legacy contamination on earth.

Case in point. While the Cold War ground on, toxic chemicals leaked from the weapons buried at Leonidovka and spread across 100 square miles of forest land—or what used to be forest land. Studies produced by independent scientists in 1997 showed arsenic levels at the Leonidovka dump surpassing 15,000 times the maximum Soviet safety level. More daunting, however, were the above-normal concentrations of pollutants that had begun to appear in the sediment of tributaries leading to the Sursk Reservoir, nearly three miles away.

Certain dioxins cause cancer, birth defects, and a host of other awful conditions. Exposure to sufficient quantities of arsenic causes the victim to suffer bouts of vomiting and terrible pain before dying of organ failure. And so on. The scientists noted that over half a million people drew their drinking water from the Sursk. Everyone agreed that something had to be done.

But Russia's signature on the Chemical Weapons Convention gave it up to fifteen years to eliminate its chemical weapons arsenal. Fifteen years in bureaucrat-speak translates more effectively to twenty-five or thirty, especially since, as everyone knew at that time, the Russian state was essentially bankrupt. At one point General Stanislav Petrov, the commander of Russia's radiation, chemical, and biological defense troops, went on record as saying it would take $5.5 billion to dispose of his country's stores of chemical weapons properly. Between 1996 and 1998, however, his government had committed less than four percent of this amount.

The potential for a full-on health catastrophe loomed. As could probably be expected, not a single Soviet entity nor any combination thereof stepped in to claim responsibility for this dire situation. And yet, somebody still had to clean up the mess. The question was: Who would do it?

As bad as things were at Leonidovka, conditions were much, much worse about 200 miles north, in the city of Dzerzhinsk.

Located on the Oka River in Nizhny Novgorod Oblast, the city had played host to the Soviet Union's primary chemical weapons plants for over half a century. The factories at Dzerzhinsk produced Sarin, VX, and mustard gas, as well as chemicals such as methylamine, phenol, and nitrobenzene, which serve as components for all manner of munitions, resins, and explosives for battle, industrial purposes, and so on. The Soviets considered Dzerzhinsk so pivotal to their national security that, until very recently, foreigners were denied entrance to the city.

According to official Soviet sources, Dzerzhinsk stopped manufacturing chemical weapons in 1965. Even so, the city still maintains some thirty-eight factories engaged in producing thousands of chemical products, which they export across the globe. And it's clear that the damage is already done. Facilities in Dzerzhinsk had a habit of dumping their effluents improperly, creating the region's most prominent landmark, a wasteland called the "White Sea."

Picture a lake of highly toxic sludge measuring 100 acres wide. Its muddy shores have been buried under rotting metal storage barrels discarded like broken toys. For more than half a century, the deadly contents of this lake have infiltrated Dzerzhinsk's surface and ground water tables. As reported in a survey by *Time* magazine of the world's most polluted places, Dzerzhinsk's own environmental agency released a startling figure: nearly 300,000 tons of dangerous neurotoxins have poisoned three generations of the city's residents. Water samples taken within the city limits show that dioxin and phenol amounts have proliferated to thousands of times above recommended levels. Even veterans of the pollution agenda find this situation hard to believe.

In 2003, the death rate in Dzerzhinsk exceeded its birth rate by a staggering 260 percent. Nearly one quarter of all deaths and more than eighty percent of all reported illnesses have been attributed to environmental contamination. According to

Greenpeace, the average life expectancy of city residents shrank to forty-five years—the lowest in the world, roughly akin to that of a human being living as a hunter-gatherer about 40,000 years ago. The situation was so bad that *Guinness World Records* called Dzerzhinsk "the Most Chemically Polluted City on Earth."

Bottom line: Cold War weapons are the filthy gift that keeps on giving. Someone has to clean them up before future generations suffer the sin supplied by our past.

CHAPTER SEVEN

Miners Big and Small

Hendra Aquan, Pure Earth's program assistant in Jakarta, stared out the window as our plane low-banked over the jungle. I craned my neck to look past him and saw the emerald carpet far below, stretched as far as I could see in both directions, its deep-pile plushness broken occasionally by homesteads or, here and there, the looping blue umbilicus of the ancient Kahayan River.

Even from such a high altitude, the rain forests of Borneo lived up to their reputation. At 130 million years old, they are the oldest of their kind on earth, older than the Amazon and the Daintree Rainforest of Queensland, Australia. Humans have nibbled their edges, but the vast majority of Borneo's jungles remain shrouded in mist, uncharted and virtually impassable. Orangutans live there, along with the proboscis monkey, pygmy elephants, Bornean rhinoceroses, the reclusive clouded leopard, and an estimated 15,000 species of plants, many of which have never been properly studied.

Our plane landed on the outskirts of Palangkaraya, the capital city of Central Kalimantan Province. The airport looked like it belonged on a sixty-year-old postcard from the tropics. Its buildings' white walls and red-tiled roofs owed much to the double-peaked, boat-shaped *jabu* structures of North Sumatra. There was only one terminal, and there we met our local guide, Pak Kulansi, as dark and brooding a figure as his headhunter forebears. Using more gestures than words, he packed us into his Xenia minivan, a tiny Japanese toaster on wheels, and drove us out to see the damage for ourselves.

We zoomed down a road buttressed by thick-trunked trees with gorgeous sweeping boughs. The leaves were so green, I thought I'd never seen the color before; so green, I felt that if I touched them, my fingers would come away wet with paint. There were towering coconut, mango, sungkai, rambutan, and durian trees. They grew so thick by the sides of the road that they seemed like the walls of a tunnel leading us forward toward a badge of light.

With Hendra translating, Pak Kulansi apologized. Evidently, we had arrived too late to sample the fabled mangosteen fruit whose white flesh, pulled from its soft purple shell, spoils too quickly to find it regularly outside Southeast Asia.

I was about to comment on the scene's beauty when the trees disappeared all at once and the blacktop vanished from under our tires. The Xenia kept moving, its engine growling in distress, but the forest had been scoured away. We were bumping through a desert of long, rolling hills whose sand had been bleached to a sickly gray pallor. It was a sterile place. The only visible forms of life were patches of scrub grass, their blades straining to reach the sun, and the trunks of limbless trees that stabbed like spears toward the deep blue sky.

I rolled down my window and gaped at the scene as Pak Kulansi slowed the Xenia to about five kilometers per hour. Turning off what remained of the road, he steered us into the heart of the wasteland. Gravel popped under our tires as he swung the minivan's wheel. We weaved around gullies and skirted arroyos, the Xenia less like a minivan now and more like some prototype moon rover, bouncing along through fields of stone.

Everything felt eerie and still. More than once, we swerved around rocks the color and size of an elephant's skull.

"What is this place?" I asked.

No one answered me, but then the pits began to appear, craters the length of Olympic-sized swimming pools gouged some three stories deep in the earth. Some were dry, but most were filled with stagnant, murky liquid.

Hendra pointed to one pool and said, "The rainy season fills them. This is water mixed with mercury."

Pak Kulansi muttered something. Hendra nodded and translated.

"He says that we cannot recover this land. The shores of the rivers, he says, are worse."

"There must be *some* way to fix it," I said.

Hendra frowned. "Maybe nature will repair the damage if people leave this place alone. But it will take a thousand years."

The wasteland I was staring at had been caused by artisanal small-scale gold mining, or ASGM. The practice might sound quaint at first. After all, we prize artisanal cheeses for their terroir, and small-scale brewers make interesting beer. Unfortunately, that's where the feel-good vibe ends.

According to some recent estimates, artisanal small-scale mining provides the basic livelihood for more than fifteen million people all over the world. These are some of the poorest people on earth who, with no other means to provide for themselves or their families, extract gold from the ground using toxic mercury. And, of course, they take no precautions while doing so—not for their own health, nor the environment's.

On Borneo, the problem started about forty years ago when the island's native Dayak tribes discovered that hunting for gold could be much more lucrative than subsistence agriculture. So they took to the rivers and streams in droves, and started to pan. It was hard work, but profitable. Most local miners brought home enough gold to keep them out of the fields for good, but their enterprise changed drastically when a large-scale mining concern moved in.

That company's name was PT Ampalit Mas Perdana. During its years in Central Kalimantan, it extracted gold from an alluvial source, which is a pleasant way of saying that it dug up the rainforest to get at the ore-rich sands beneath. Like all modern miners, PT Ampalit processed these sands using chemicals to extract the desired mineral content. By and large, however—like all modern miners—the company used best practices. The upshot was this: PT Ampalit employed thousands of workers and destroyed several thousand hectares of ancient rainforest before going broke in the Asian currency crisis of the late 1990s.

Which wasn't the end of the story at all. If anything, it was the beginning.

With PT Ampalit out of the picture, the locals were free to work almost wherever they cared to. They had spent years watching PT Ampalit's methods, and they'd figured out that panning is an inefficient method to hunt for gold; it's labor-intensive, with very low yields. So they took a page from PT Ampalit's book and created their own alluvial extraction method. It was a primitive process to say the least, a bastardized version of what the better-heeled and more technologically sophisticated big company had practiced.

Though they didn't know it at the time, the small-scale miners' method was an ancient one, dating back to the sixteenth century. It involved the uncontrolled use of a highly toxic element.

Mercury.

The Xenia bounced past the skeletal frames of small huts crumbling to dust in the sun. According to Pak Kulansi, these were once the miners' homes. Through Hendra, he described how the miners construct temporary villages, building "houses" by wrapping tarpaper around hastily constructed wooden frames and tearing them down whenever the gold runs out. At which point they simply move to another patch of forest and repeat the process all over again, leaving desolation in their wake.

We turned up a stream bed pressed into service as a makeshift road. Pak Kulansi steered us toward a cluster of huts with sagging, bright blue tarpaulin roofs. Chickens strutted and bobbed before our wheels, their postures drooping a bit in the heat. We passed a massive, rusting satellite dish, as incongruous here as a clown on the moon.

In the center of the dead village, Pak Kulansi parked and got out. Hendra and I followed suit and trailed behind him up the main thoroughfare, if you could call it that. Everything was intensely quiet. We heard nothing but the crunch of our boots over barren crust—until the chickens squawked, a sound so harsh it nearly made me jump out of my skin.

At the very edge of town, near the bounds of the wasteland, Pak Kulansi stopped and gestured. Half a kilometer from where we stood, across the chessboard of pits in the earth, an ancient forest stood proud and tall, its border as abrupt as the face of a cliff. A wall of towering trunks supported a green canopy so dense that it blotted the sun. It was a beautiful sight, and one that immediately fired my imagination. I could picture everything under those sweeping high boughs sleeping cool and wet in a darkness as deep and hallowed as time.

The contrast between the two landscapes came as a huge shock. Waste against jungle. Death against life. The scrub I was standing on—this bald scrabble of earth spread over several kilometers—had once been forest, too. Strong and majestic. Now it was cracked plate, growing fast, and with nothing to stop its advance.

Pak Kulansi pointed to the forest and spoke one word, which Hendra translated.

"Miners."

Then he put one hand to his ear. He motioned that Hendra and I should do the same. And that's when I heard it, the distant growling sound.

Chainsaws.

In the distance, tiny smoke puffs appeared. At the edge of the forest, a tall tree shuddered, tipped, and fell.

Pak Kulansi stared at the woods, his jaw set hard in a line.

◆

This is how the process works. First, the miners clear-cut patches of forest. The woods they destroy are often home to the orangutan, a highly endangered species. With dwindling numbers of trees to support them, the orangutans die of starvation and exposure. Some get caught in the open, where the miners shoot them on sight for food.

Once the trees are down, the miners bring in diesel generators to pump high-pressure jets of water at the exposed root systems. The water washes away the stumps along with all the alluvial sand, leaving pits in the earth like the ones we had seen. The miners then suck up the mud using another set of generators and pass it through long sluice boxes equipped with special mats that scrub out the larger, mineralized ore particles.

At the end of each day, the miners dump these ore particles into pails, to which they add generous portions of liquid mercury. For several minutes, they squat on their hunkers, stirring the toxic mixture by hand. As I mentioned before, this is an ancient method of mining. Prospectors of centuries past learned what these Dayak miners know quite well: mercury bonds with gold to create a lump of gray amalgam. This allows a miner to capture a great deal more gold than he could by hand. Even the finest gold particles bond with the mercury, creating a small silvery lump that runs as big as a robin's egg in most cases. The miners collect these one-centimeter balls in parachute silk, wrap them tightly, and tie off the ends before sticking the lumps in their pockets.

Final step: the miners take their little gray balls to dealers at a *toko emas*, or gold shop, one of many such enterprises sprouting in Borneo's boomtowns. Using blowtorches operated by foot pedals, the gold shop owners burn the mercury out of the tiny lumps and voilà! What's left is a nugget of nearly pure gold that fetches a handsome price in cash. A win for the miners. Or is it?

Unfortunately, torching the lumps of amalgam creates mercury vapor, which disperses into the air, where it eventually condenses on the walls of the shop, in nearby deposits of soil, in water—anywhere. Exposure to mercury blocks the phosphate necessary to replicate DNA, which in turn destroys neuron pathways. It can lead to respiratory and cardiovascular disease, psychotic reactions, miscarriages, and death. This is tragic when it happens to an adult. When it occurs in children, the results can be awful. Exposing children and fetuses to mercury permanently damages their developing brains.

Mercury poisoning was glaringly apparent when we met the owner/operators of a toko emas. Because of prolonged exposure to mercury vapor, their eyes were bright red and slightly bugged out, their movements jerky and spastic. They suffered from the same malady that had afflicted so many haberdashers in Victorian England. In those days, hat-makers used to stiffen the felt of the hats they were making by brushing it with liquid mercury. Mercury poisoning became their occupational hazard, a fact people openly acknowledged and which Lewis Carroll drew from when he created the Mad Hatter character for his book *Alice's Adventures in Wonderland*. Recently, however, thanks to public information campaigns (many of which were ultimately sponsored by Pure Earth), the toko owners have learned to protect themselves, their families, and their employees.

Pak Kulansi took us to see a toko emas in Kereng Pangi, a hardscrabble town that sprang up from the dust about twenty years back. Tie some horses to hitching posts, film the main thoroughfare in black-and-white, and Kereng Pangi would almost have made the perfect set for a Hollywood Western. The atmosphere was brittle and hot, and most of the roads were hard-packed earth. Above them hummed the power grid, a network of sagging black wires crisscrossing atop jackleg poles. The neighborhood mosque was a modest place, a couple of stories tall but, for all that, the tallest structure in town. Loudspeakers hanging from its cornices sang a tinny-toned summons to afternoon prayer.

The kid who ran the toko we entered looked to be about fifteen years old, but he wasn't. Pak Kulanksi introduced him as Pak Rizky, who then waved us inside and offered us cold cans of Sprite before proudly giving us a tour of his enterprise. First, the blowtorch and crucible. Then his sets of scales. Finally, his cases overflowing with gold.

"This shop does well," Pak Kulansi said.

"I'll bet," I said.

Three glass cases were crammed with gold earrings, necklaces, bracelets, and charms. It boggled my mind to estimate the worth of what I was looking at. Then I thought of the price.

Destroying the Bornean jungle is the most obvious environmental damage caused by artisanal small-scale gold mining on Kalimantan. Less obvious, though no less dangerous, are the significant quantities of mercury that escape as the miners and gold shop owners work. This mercury pollutes the local environment and spreads in ways that are fairly insidious. For example, in early August 2014, the *Jakarta Post* reported that thousands of dead fish were found floating, belly up, in the Meriam and Teunom Rivers in Aceh Province on the big island of Sumatra. The likely cause? Mercury used in artisanal small-scale gold mining.

Since 2009, some 250,000 previously protected hectares near the two rivers have attracted both legal and illegal mining activities. Mercury and other chemicals used for mining have apparently infiltrated the water supply and floated downriver. The dead fish found that August were dissected by marine biologists. Each fish had swollen internal organs and had suffered internal bleeding. Everyone's heard the old adage "Give a man a fish and you'll feed him for a day; teach a man to fish and you'll feed him for a lifetime." The newer adage you'll hear in these regions is "Give a man a fish and you'll feed him for a day; teach a man to fish and he'll be dead in five years from mercury poisoning." These fish were also a staple for the local population, which introduces the next leg of the crisis.

Scientists are still researching mercury's behavior, but preliminary observations indicate the following chain of events: vaporized mercury enters the atmosphere, precipitates, and settles into water supplies, where it mingles with anaerobic sediments that bacteria consume and convert into the potent neurotoxin, methyl mercury. This is where the crisis goes global.

Mercury swept downstream in Indonesia gets absorbed by kale and microorganisms that live in silt layers on the bottoms of rivers and oceans. The marine microorganisms slowly convert the mercury to methyl mercury. Small fish eat these microbes and get contaminated. The small fish get eaten by larger fish, which in turn get eaten by larger fish, which eventually land on our plates. Which is why, these days, many Westerners take it for granted that certain fish transmit mercury when eaten. Pregnant women have become especially wary, and many obstetricians tell their expecting patients to avoid fish altogether.

A twenty-year lag seems to exist between mercury's release and its appearance in the food supply, but it gets there eventually, and in ever-increasing proportions. The sad truth is that the mercury we saw being released in Kalimantan can end up in the tuna fish sandwich our children's kids will eat at school someday.

Westerners are generally aware that mercury is a problem in some fish species, but few people understand the source of the contamination. The majority of the mercury showing up in our food—more than half of all quantities found, according to some scientists—originates from ASGM activities around the world.

Studies estimate that ASGM activities in Indonesia release about 200 tons of mercury into the environment each year, an astonishing quantity. But that's nothing compared to the rest of the planet. In the past decade, the practice of ASGM worldwide has gone parabolic. The United Nations Environment Programme estimates that up to twenty million people in at least seventy low-to-middle-income countries host ASGM activities. Nigeria, Senegal, Zimbabwe, Tanzania, China, Laos, Brazil—the list goes on and on. So while Indonesia releases 200 tons of mercury a year, the rest of the world releases almost seven times that amount.

So how do we deal with this problem? The answer, in my opinion, is carefully. Because while some solutions offer quick fixes, they actually create more problems down the road.

For instance, in the summer of 2014, the *Washington Post* reported that Peru had begun helicoptering police and soldiers into ASGM mining camps in the country's Madre de Dios region and blowing them up with explosives. In recent years, the article states, some 40,000 miners have laid waste to more than 120,000 acres of prime Amazon rainforest in almost precisely the same manner we witnessed being practiced in Central Kalimantan. Chainsaws. Generators. Water jets. Sluice boxes.

And mercury. Experts estimate that these miners release thirty to forty tons of mercury into the environment each year. The pollution generated by these quantities creates the same kind of damage that the *Jakarta Post* reported was taking place in the rivers of Sumatra.

The Peruvian government tolerated ASGM activities for

years. Standards and enforcement lapsed while the miners grew more numerous and sophisticated by almost exponential degrees. Now that the miners are running the show, the Peruvian government feels that it has to stamp its foot down. Hence the militarized solution.

"These people have done extraordinary damage," Antonio Fernandez told the *Washington Post.* The article describes Fernandez as "Peru's top prosecutor for environmental crimes." Dressed in military fatigues and combat boots, he watched a strike force blow up a village of miners before consenting to be interviewed. "We have to respond with the same amount of force."

Fernandez's stated goal—indeed, the goal of the Peruvian government—is eradication. They don't want to work with the miners. They want to rout them from the region.

"They don't belong here," Fernandez told the *Post* as the village burned around him. "They should go home."

Which, of course, is the root of the problem.

I wasn't surprised to read the *Post* article's description of the miners in Madre de Dios: mud-caked tribesmen, mostly poor, who hailed from the country's wild interior. Who else would commit to a life like that? ASGM is not a middle-class industry. Like the Dayaks of Borneo, Peruvian Indians view digging up rainforests and sluicing gold from the dirt as their only chance to better themselves economically.

Are their actions inappropriate? Sure. They're poisoning themselves and the environment. But we can hardly blame them for doing what almost every other underprivileged person on earth hopes to do: create better opportunities for themselves and their families through any means possible.

Blowing people up doesn't stop them from being poor. In the long run, violence solves nothing. We need to find better solutions, and who better to do this than government? Indeed, what else is a government for?

Why not invest in new industries that set the trend for sustainable practices? Peru is a nation endowed with vast and impressive natural resources. It is the government's responsibility to properly manage those resources. Surely there are ways to help these locals participate in other non-damaging industries, ways that can help them feed their families while maintaining the forests. The focus should be on finding ways to make a business out of protecting those great resources. Why not sponsor training programs geared at preparing unskilled laborers to work in these new sustainable industries?

For example, at present, Peru is the world's fifth largest exporter of gold. Most of this gold gets mined by large multinational conglomerates that operate within strict sustainability guidelines. Therefore, why not prepare and refocus the miners to serve in industries that operate with everyone's best interest in mind? Peru could make it an operational condition that multinationals accept certain numbers of local workers. Train them, employ them, give them a say in their own affairs, and let them start building their own presence in the world.

Programs of this type have worked extraordinarily well in Central Kalimantan, where large-scale mining companies employ local Dayaks as laborers, guides, surveyors, translators, and cultural attachés. The trick is to practice the politics of inclusion. Let everyone contribute to the region's success and profit in a manner that is personally, commercially, and ecologically self-perpetuating.

These, of course, are mid-term goals. Immediately, something has to be done about the persistent release of mercury at ASGM sites worldwide. Pure Earth has been monitoring a region of the Philippines where miners extract gold without mercury, and do so more efficiently. Their technique employs the concentrating techniques of sluice boxes with borax added at the final stage. No mercury.

A chemical ingredient common to many household and

commercial products, borax is technically a salt of the element boron. As chemicals go, borax is generally benign. Occasionally, borax dusts and powders can irritate the skin and mucous membranes of people who handle them, but only in large amounts, and the irritation has no permanent impact on health. For this simple reason, borax is, overall, a million times safer for ASGM activities than mercury ever will be.

This is now called the Philippines Method (and sometimes the Borax Method), since it was first developed about three decades ago near Luzon. Specialists, however, tend to refer to it according to the key physical property the method employs, which is gravimetric separation. Basically, gold ore runs through a sluicing stage, the same as in mercury-based ASGM. The result is the same: a final, concentrated slurry that yields a higher percentage of gold. After that, it is panned to concentrate it further. But what happens when you add borax? The element lowers the yellow metal's melting point, which in turn enhances the quantity and quality of gold recovered.

That's good news for the miners. The Philippines Method represents a technological upgrade for them, a means by which they can recover more and better gold without the involvement of expensive, toxic elements. As of the writing of this book, Pure Earth has staged successful interventions of this sort in Bolivia, Mongolia, Peru, and parts of Indonesia. By successful, I mean that miners have switched practices both enthusiastically and profitably while eliminating toxic output from their work.

There are often wrinkles. For instance, different ores require adding more complexities into the process, and often it takes a while to get the formula right when introducing it to a new region. And in some places, miners who are taught the new method don't want to share it with others.

But we know that miners will switch to the Philippines Method (or any method, for that matter) when they see that it represents a clear, immediate, and cost-efficient technological upgrade to their practice. The emphasis here is on "cost-efficient." Enhanced income seems to represent the basic motivating force in almost everything the miners do. Put differently, if increased profit isn't part of the matrix, our chances of grabbing the miners' attention are slim to none. This makes sense when you consider that almost every ASGM miner lives hand-to-mouth. They don't have the wiggle room to pause their enterprise and tinker with some new technology, hoping it will work, even when they know that something they're currently doing is hurting themselves and their families.

The Philippines Method is not foolproof. We've found that gravimetric separation works best in instances where the gold particles present in sluiced or concentrated ore are particularly large, that is to say, visible to the naked eye. Smaller particles simply do not respond as well to the use of borax as a catalyst or "flux." Which means that borax makes for a viable alternative in regions where, geologically, ore deposits are large and coarse. In other regions, we have to go back to the drawing board and come up with another option or options.

The U.S. and EU have already placed restrictions on mercury's export, which should reduce the element's availability while raising its price. That's one way to throw a shoe in the gears, but it won't work well past a certain point. Frankly, if we're going to gain any traction at all against ASGM, we have to take the various small-scale mining communities by the hand and do this phase of the work for them. Specifically, governments and NGOs must take the lead and actively promote and enforce the groundbreaking provisions set forth by the United Nations' Minamata Convention established in October 2013.

I won't go into all the theories backing the need for that treaty. We've already explored the matter here to a certain degree, and the treaty itself is readily available online if you're curious enough to read it. In actual practice, the convention will be best served if we acknowledge that one size doesn't fit all when it comes to small-scale miners or the methods they employ. And, since each ASGM community is different, we should develop and prove a basket of alternative methods to keep them, the environment, and the world at large healthy.

Once we have these tools in place, we can gradually limit if not eliminate the release of toxic chemicals through small-scale mining. At which point, with an eye toward preserving the environment, we can begin to focus on how or if we should allow the miners to continue their enterprises.

Again, speaking frankly, I don't see outright elimination of their sector as a viable possibility. It goes against everything I've witnessed regarding the behavior of economically disenfranchised people the world over. People will do what they must to survive. Trying to stop them from doing that is like beating a hungry dog with a stick. Far better to simply feed him some meat and thereby make a friend. For this and other reasons, Pure Earth and its partners advocate folding ASGM communities worldwide into established, tax- and royalty-paying, law-abiding and environmentally conscious mining firms.

This may be easier to do than it sounds. At this point, many governments and established mining operations have studied ASGM activity. They recognize the potential profits to be made by recovering resources now relegated to ASGM activity, and they can do so in a sustainable manner.

But possibilities such as this lie much farther down the road. For now, we need to concentrate on coming up with more non-toxic ASGM practices. It's the best and most immediate way to create lasting, worldwide change.

CHAPTER EIGHT

Coffins Filled with Pesticides

Of all the toxic substances Pure Earth deals with, pesticides are one that we humans release into our environment because we think they will do some good.

A pesticide's purpose is to kill living things, be they weeds, insects, fungi, or rodents. Without question, the use of pesticides has allowed the human civilization to thrive and prosper in recent decades. Just as clearly, however, the abuse of pesticides has created some astonishingly dangerous situations.

The Azerbaijan Republic is an ancient, mountainous country between Russia and Iran, bordered by Armenia to the west and the Caspian Sea to the east. A good fifty-eight percent of its terrain—some 4.8 million hectares—has been dedicated to agriculture. Azerbaijan grows cotton, tobacco, grapes, vegetables, and citrus fruit while maintaining strong output of livestock and dairy products. With so much farmland in constant cultivation, Azerbaijan requires massive quantities of pesticides. The country produced nearly half a million tons of DDT between 1958 and 1989.

For those unfamiliar with it, DDT is a colorless, tasteless, nearly odorless crystalline solid that was developed during the 1940s as the first modern synthetic insecticide. During the 1950s, countries all over the world applied regular and liberal doses of DDT to crops and regions infested by mosquitoes. But this all changed in 1962 with the release of Rachel Carson's groundbreaking book *Silent Spring*.

Carson began her career in the U.S. Fish and Wildlife Service. Her first three books explored oceanography and showed her unique skill at conveying both her education and passion as a naturalist to a public audience. During the 1940s, she began tracking the development and use of synthetic pesticides, many of which were developed by the U.S. military, and many of which showed early signs of harming the environment.

Her concerns peaked in 1957 when the Department of Agriculture requested $2.5 million to eradicate the gypsy moth, an insect whose long-haired, bumpy-backed caterpillars were feasting on the leaves of hardwood trees, defoliating whole forests across the northeast.

The plan called for crop-dusters to spray a mixture of DDT and petroleum oil across three million acres. Problems arose almost at once. Residents of Long Island, New York, filed a federal suit claiming the spray used by the USDA was killing local wildlife and crops. Milk from area cows had to be pulled from the market because it contained DDT at unprecedented levels.

Yet federal courts dismissed the suit. Carson and several of her colleagues at the FWS took umbrage, as did members of the Audubon Society which, at that point, was more or less a social club for affluent bird watchers. The crop dusters' spray killed thousands of birds and galvanized the Audubon Society into the force it is today.

One year later, in 1958, Congress approved another eradication program, this time to fight the supposed scourge posed by the Mexican fire ant. I say "supposed" because several reports surfaced presenting sound scientific evidence that fire ants weren't really harmful at all. Yes, they built high mounds in the middle of fields and yes, their bites could irritate skin. But a scourge? Not really.

As further research showed, the notion that fire ants were dangerous sprang largely from boogeyman stories, many of which were rather colorful. In some, the ruby red ants killed children who wandered out into local fields. In others, the ants organized raids on livestock paddocks, felling calves and piglets while horrified farmers looked on.

In truth, no proof existed to show that fire ants posed any danger whatsoever to crops, land values, or people. Regardless, Congress authorized $2.4 million for the eradication program. According to critics, this decision had more to do with appeasing certain farmers with influential connections in the House and Senate than any real concern about the state of America's crops.

Mind you, the uproar over the program to eradicate gypsy moths was still fresh in people's minds. In order to calm public sentiment, government scientists made assurances that wildlife across the country would be protected from damage during the fire ant program.

Under the fire ant program, twenty million acres of prime southern farmland were seeded with heptachlor pills. A persistent pesticide, heptachlor does not break down and dissipate harmlessly into the environment. Rather, it bioaccumulates, working its way up the food chain as smaller species consume the pills and larger species, in turn, consume them.

By some estimates, heptachlor is over thirty times more toxic than DDT. No wonder federal scientists began discovering the bodies of dead animals soon after distributing pellets in parts of Georgia. In Alabama, farmers were drawn to the smell of dead animals decomposing in heaps along fencerows.

Federal employees scurried to cover up the damage they had caused; they failed. Moreover, when inquiries revealed their attempts to deceive, the public reinvigorated its dissent to pieces of federal legislation that affect the environment.

The pot continued to stir in 1959, when the Audubon Society released its own research. A study it had commissioned detailed how, in Texas, chemicals used in the fire ant eradication program had nearly wiped out an entire species of bird. On the heels of this, more reports poured in from states like Louisiana where, in one region, the population of red-winged blackbirds dropped from 135 to zero in two months. Autopsies showed massive heptachlor concentrations in the bodies of dead mammals across the American South.

Results of this nature stirred more conservation groups to action. Organizations like the Sierra Club and the National Wildlife Federation joined the U.S. FWS and FDA in criticizing the USDA. And with that, the pesticide war was on.

Enter Rachel Carson. *Silent Spring* covered all this territory and more as a prelude to her assault on DDT. According to Carson's research, DDT, once hailed as a miracle of modern science, was actually a toxin with a nasty penchant for bioaccumulating in mammalian fatty tissues. The pollution it caused was so extensive, in fact, that human mothers had inadvertently breastfed trace amounts of DDT to their babies.

Later research conducted by independent organizations confirmed this supposition, as well as the fact that, in human beings, DDT can cause increased rates of cancer and other diseases. For some animals, the stuff kills outright, even in miniscule amounts. Widespread use of DDT in the United States has subsequently been linked to vanishing bird populations, including endangered species such as the peregrine falcon and the bald eagle.

Carson endured a great deal of criticism on behalf of *Silent Spring*. Scientists on a government review panel for pesticides accused her of shilling for the environmental movement. Carson rebuked them. Many of her accusers, she said, had ties to industries destined to profit, should pesticides proliferate. It's worth noting that even Carson's staunchest critics admitted that pesticides pose potential risks to human health. This, plus the groundswell of public interest caused by *Silent Spring* led to an outright countrywide ban on DDT in 1972.

A few years later, the U.S. EPA led a global initiative to control the use of DDT and other persistent pesticides worldwide. This might have made a pleasant coda to a potentially tragic tale, but it didn't play out like that in most poor countries.

Between 1989 and 1990, Azerbaijan joined the growing international movement to ban DDT and similar toxic pesticides. But what could it do with the massive stockpiles of the stuff it had already generated? Answer: seal the toxins in metal "coffins," drive them fifty-three kilometers outside Azerbaijan's capital city of Baku, and bury them underground in so-called "pesticide cemeteries." In this fashion, they could no longer do any harm.

At any rate, that was the notion.

Years passed. Corrosion set in. The metal coffins began to leak. As of the writing of this chapter, the poisons contained in those sarcophagi have infiltrated local soil and water tables, causing contamination across several strata.

The chemicals have crippled many species of wildlife. Local populations of the endangered Caspian seal have been affected. The stock of sturgeon, a staple food and one of Azerbaijan's key exports, has declined. Indigenous plants have withered and died. High rates of cancer have appeared in human populations living in close proximity to the pesticide cemeteries. Since these populations are mostly poor to begin with, the increased burden placed on them by disease reinforces "an irreversible effect on the socio-economic [disadvantage] in the country," according to the world-renowned Lancaster Environment Centre based in the UK.

Azerbaijan reports that its stockpiles include approximately 4,000 tons of obsolete pesticides. In 2006, its government adopted a plan to improve the country's environmental outlook. They could use some help, to say the least. The country's pesticide tombs are deteriorating quickly and will soon create an even more widespread disaster. A recent study co-financed by the University of Applied Sciences of Northern Switzerland highlighted the need for "fast follow-up action" at the sites near Baku. The UAS's advice aptly extends to similar crises found throughout the territories of the former Soviet Union.

Armenia. Belarus. Georgia. Kazakhstan. Kyrgyzstan. Tajikistan. The list goes on and on.

Is there hope? Of course there is, though hope alone won't turn the tide. It's time to put our money and our muscle where our mouths are. Fortunately, some very smart people are doing exactly that.

In September 2011, following the 11th International HCH and Pesticides Forum, the European Union announced a project dedicated to "improving capacities to eliminate and prevent recurrence of obsolete pesticides as a model for tackling unused hazardous chemicals in the former Soviet Union." The EU wasn't just talking a good game; it offered a grant of €7 million to get the ball rolling. Estimates state that a comprehensive cleanup for the entire former Soviet region will cost 100 times that amount. But hey. You've got to start somewhere.

The fact that a forum was named (partially) after HCH strikes me as telling. It signifies that European officials respect the breadth and scope of the pesticide problem. Importantly, they also recognize its subtlety.

HCH is an acronym that stands for hexachlorocyclohexane. If you can't pronounce the name of the compound, I understand—I have trouble with it, too. Just know that HCH leads a pack of pesticides we no longer use, and not because they kill outright, but because their interaction with the environment is suspect.

In my opinion, this is precisely where the battle lines need to be drawn. Obviously toxic materials should never be used—that's a no-brainer. But nor should we tolerate the use of those pesticides whose interactions with the environment have not been clearly, broadly, and intensely vetted. HCH falls in this category.

Originally synthesized in 1845, the compound's pesticide applications weren't discovered until nearly 100 years later. Products spun off from HCH can destroy a broad range of insect life, including fleas, ticks, lice, and leaf-destroying worms. Hoping to capitalize on these applications, a UK-based company began marketing a form of the compound (gamma-hexachlorocyclohexane) as the product lindane.

Lindane found widespread use as a spray treatment for seeds, soil, and crops, as well as an anti-pest dip for pets and livestock, and also as a household insecticide. Pharmaceutical applications evolved. Lindane-based lotions and shampoos showed tremendous efficacy at treating cases of lice and scabies.

The product was deemed so effective, in fact, that over the next fifty to sixty years, approximately 600,000 tons of HCH were produced by major industrial countries all over the world. The U.S., Brazil, China, India, Russia, several European countries . . . everyone used lindane. Until the mid-'80s, when a number of sobering health reports began to emerge.

Lab rodents that were fed lindane over prolonged intervals began to display an elevated risk for liver and thyroid cancers. Use of the product fell under intense scrutiny. Lindane's proponents argued that the medical reports were inconclusive. Fears based on lindane's use were misleading, they said. For instance, human beings would have to digest an enormous amount of the product over unusually long periods to put themselves at risk.

Lindane proponents also noted the infrequency of cases where human beings exposed to lindane had contracted cancer. They also argued that, in the few cases where cancers had developed, those stricken had also been exposed to many other harmful chemicals.

For instance, one study conducted in Minnesota and Iowa showed that male farmers who had handled lindane showed an elevated incidence rate for non-Hodgkin's lymphoma. But these farmers had also handled other pesticides. Why should lindane be seen as the culprit when other agents could just as easily have been responsible?

Indeed, a later study conducted in the American Midwest seemed to confirm that lindane played no major role in contracting non-Hodgkin's lymphoma. And another inquiry seeking to prove that lindane shampoo caused childhood brain cancer returned no hard evidence that the product played any role in such pathology.

But health activists countered that certain occupations fit the high-risk category perfectly. If, as the experiments with rodents seemed to show, massive and prolonged exposure to lindane created a risk of disease, then surely farmers stood right in the crosshairs; they used lindane every day, and many did so over the course of their entire careers.

Farmers, of course, weren't the only occupations at risk. What about people employed as crop dusters, sheep ranchers, pet groomers, and veterinarians?

For that matter, what about children?

Everyone heaved a sigh of relief when studies could not produce hard links between certain cancers and the use of lindane anti-lice and scabies shampoos. But where exactly do we draw the line? Should children be given shampoos or lotions that contain chemicals *suspected* of being harmful? Shouldn't we be alarmed at the fact that lindane made its way into the breast milk of mothers who'd been exposed to it? Granted, lindane wasn't nearly as toxic as DDT or other pesticide compounds. But who could endorse the product's use if even mild risks were suspected?

Situations like this one define the precarious balance that exists when dealing with pesticides as a social force. On one hand, they deliver great economic benefits. Crops free from pests yield in greater abundance. In industrialized countries, this makes food more available and more affordable while greatly reducing the likelihood of famine and other agro-related catastrophes.

The trick, of course, lies in finding a way to enjoy the benefits pesticides offer while minimizing if not entirely negating the damage they cause to human health and the environment.

This is easier said than done.

Consider a study released in 2013 that outlined a link between the development of endometriosis and exposure to beta-hexachlorocyclohexane, a byproduct of lindane's chemical synthesis. Imagine the can of worms this opened. Even supposing that lindane could be used without injuring people or polluting the environment, you can't produce the pesticide without also producing beta-HCH, concentrations of which have turned up in water and soil samples all over the world.

In 2005, officials from an Italian governmental agency tested milk from dairy cows in one of their country's central provinces. The milk, they found, contained beta-HCH concentrations that were twenty times greater than the maximum legal mandate. That's especially bad news since beta-HCH is a proven neurotoxin. Exposure to it can damage our brains, causing Parkinson's and Alzheimer's disease among many other afflictions.

So again, it's not such a cut-and-dried issue. The environmental footprint left by pesticides doesn't always correspond to the most obvious vector. How often do sprayed pesticides drift on the wind and settle in local water sources? How do the chemicals used in pesticides interact with chemicals used in other industrial processes? How long do pesticides and their byproducts linger in our environment, and in what form? What about the effluents created from the manufacture of pesticides? And so on.

For these and other reasons, fifty-two countries opted to ban the use of lindane in 2006 (another thirty-three countries called for restrictions on the product's application). Use of such products has declined as more and more nations take vocal and adamant stances against them. Which brings us back to the matter at hand:

Regardless of current controls, pesticides still comprise one of the world's most virulent and daunting forms of pollution. At present, the U.S. EPA acknowledges some seventy chemicals that, while used as key ingredients in pesticides currently on the market, are also known to cause genetic damage leading to disease and death.

Activists charge that the labeling process for active chemical ingredients has been obscured by political pressures. Some say that lobbyists for certain corporations have co-opted the legislative process to push their products forward, though many have not undergone full testing. Clearly, this is a cause for concern. The United States is a powerful Western nation that boasts perhaps the most robust slate of environmental legislation. But if our own approach to certifying pesticides is slapdash, how can we expect things to fare in countries with weaker or sometimes nonexistent enforcement of public health codes?

The answer to this question was handed to us in the summer of 2013. On July 16 of that year, children studying at a primary school in the village of the Dharmashati Gandaman (Bihar Province, India) complained that their cafeteria lunches tasted strange. On the menu that day: soya beans, rice, and curried potatoes. The meals were part of India's national Midday Meal Scheme, a program that supplies food to nearly 120 million children—the largest of its kind in the world.

Reportedly, the headmistress at Dharmashati Gandaman primary rebuked her students for complaining. They went back to eating their lunches and, within thirty minutes, dozens of kids between four and twelve years old doubled over, stricken with diarrhea and vomiting.

They cried out in pain, sprawled in hallways, asking for aid. So many children fell ill that they overwhelmed the school's medical staff. Many were shipped out to clinics while more were sent home to their families. By the end of the day, twenty-seven kids lay dead. Sixteen expired at the school while four passed at local hospitals. The final seven expired as their parents, siblings, and neighbors tried desperately to save them.

Beyond these fatalities, forty-eight more children, drastically ill, were moved to facilities where, fortunately, doctors were able to stabilize them.

What happened?

By now, you can probably imagine a host of terrible possibilities. Because now you understand that pesticides (and especially obsolete pesticides used in developing nations) can linger in rural areas, causing environmental destruction far beyond the scope and interval of their intended use.

In the case of the Dharmashati Gandaman primary, it was children who paid the price.

Subsequent tests confirmed that a bottle of cooking oil used to prepare meals that day contained immensely poisonous quantities of monocrotophos, a cheap organophosphate used as an agro-pesticide in several developing nations, including India.

Organophosphates serve as the basis of many insecticides, herbicides, and nerve agents. The Nazis reportedly produced large quantities of organophosphate for use in chemical weapons that, thankfully, were never released before World War II drew to a close.

Since 1966, over 650 agricultural workers have been poisoned by organophosphates in the U.S. alone. Of these, 100 reportedly died. Casualty statistics of this sort are largely unavailable from developing nations, though the obvious presumption is that they're exponentially greater than their Western counterparts.

The U.S. banned the use of organophosphates in 2000. That's a step in the right direction, of course. But only one step in a much longer journey.

Another good step: at present, the use of organophosphates has been banned or restricted in twenty-three countries. More than double that number have categorized the import of organophosphates as illegal.

So yes. There is progress. It's just not enough. Nothing can return the slain children of the Dharmashati Gandaman primary to their families. But we can prevent similar instances from happening to children or other vulnerable populations across the world.

In late winter of 2014, Pure Earth won a small contract—$160,000—from the Food and Agricultural Organization of the United Nations. FAO asked us to characterize an obsolete pesticide dump in the Ahava Valley of Somaliland. In technical terms, we were asked to perform a remedial investigation: our experts would survey the site and define options for remediation based on the conditions they found present.

We knew a little about the region's profile before we went in. Specifically, we knew about locusts, which lay at the root of the pesticide problem. East Africa had its share of pests, many of which could be virulent. It might be difficult to imagine but, throughout Africa and Western Asia, swarms of locusts still have the potential to wreak the kind of wholesale destruction they did in ancient times.

Small swarms of locusts can spread across a few hundred square meters, while larger swarms fill areas more than 1,000 kilometers square. Billions of insects appear on the horizon like a fast-moving cloud, blackening the sky before landing on crops, eating everything in sight, and then departing, leaving ruin behind.

These attacks mirror those described in the Bible, the Koran, and other ancient texts: total devastation sparking famine across many nations. Accordingly, several organizations exist or have existed to combat the scourge that locusts present. One of them is the Desert Locust Control Organization of East Africa. In years past, the DLCO used the site in Ahava Valley as a storage campus for organochlorine pesticides such as DDT, aldrin, dieldrin, and lindane.

Most of these pesticides are illegal now, banned by the Stockholm Convention, which refers to them as persistent organic pollutants because they stay in the environment forever, refusing to degrade. This became especially problematic during the early days of the Somali civil war. The history of that conflict is too long and detailed to cover here. Suffice it to say that some bombs were dropped; the ordinance basically turned the storage campus to rubble and spread unquantifiable amounts of organochlorines all over the place.

As the war continued, a number of internally displaced people were encouraged to settle in the Ahava Valley, near the pesticide dump. We found them living under a cloud of awful stench, which confused us at first. The organochlorines we knew about were toxic, but they shouldn't give off that kind of an odor. So we reconnoitered the site and found the culprit: apart from the organochlorines, someone had deposited a great many drums of the organophosphate pesticide malathion on the campus.

Malathion is one of the most commonly used pesticides in the United States and around the world. It's also toxic and gives off a terrible odor. By questioning the locals, we determined that the drums had been stored there about ten years before. Somewhere over that decade, however, they'd begun to leak, and that's what created the stench. Also, the organophosphate was mixing with the organochlorines present, creating a kind of sinister cocktail. The resulting mixture—contaminated soil and leaking drums—wasn't any more deadly than the sum of its parts. But the whole scene was certainly a mess and definitely harmful to those people living nearby.

When we finished our survey, we brainstormed about immediate measures we could take to make the site more livable. We hit on the idea of purchasing some overpack drums—eighty-five-gallon barrels which, as their name probably implies, are large enough to accommodate normal fifty-five-gallon drums,

Dutch oven style. We could sheath the leaking drums of malathion in the overpack drums, seal them, and cart them away for destruction. A simple intervention that could make a big difference, and quickly. We decided we could achieve this, and more, within the budget we'd been allocated just for evaluation.

It took a while for the drums to arrive since, technically, Somaliland doesn't exist. To date, it isn't recognized by other nations and this, to say the least, confounded the import/export process. I can tell you stories that will curl your toes about how difficult it is to get anything into that country at all. It's a logistics nightmare. Just getting the overpack drums to the site was crazy, not to mention expensive. Eventually, however, we prevailed.

At the site, we set about nesting each leaking drum of organophosphate into an overpack drum. We also stuffed the overpacks with as much contaminated rubble as they could hold. Then we sealed them and shuttled them to a port, where they were placed in a shipping container and readied for export.

There were more complications, however. The first was that a high-temperature incinerator was required to properly destroy this kind of chemical. Unfortunately, there are few facilities of this type, and the only ones that accept waste from foreign nations are in Europe or Japan. It took a lot of paperwork, but eventually, we were cleared to send our organophosphates to a qualified facility.

The second complication arose when we tried to export a container filled with harmful chemicals from a country that doesn't officially exist. Again, it took some wrangling, lots of patience, and improvisation. But once again, we prevailed.

The first and most immediate effect at the site was a dissipation of the awful scent. The community was thrilled with us for that. But the organochlorines were still in the soil, still presenting a risk to human health. They required more complicated means to remediate. This meant a second trip, so we returned to the site and mixed a quantity of soil amendments into the contaminated land. These were products designed to break down the organochlorines, effectively neutralizing them.

We view this sort of work as a safeguarding measure since, realistically, we don't have the funds to pursue a more complete intervention, nor do we think that any other organization has this particular site on their radar. Soil amendments are a well-tested solution in the U.S. They seemed to have worked well in this instance, too. The levels of pesticides in these soils are no longer a worry.

At the risk of tooting our own horn, the actions we took were over and above the scope of work we were contracted to perform. We ended up completing the job on a budget that was supposed to cover only a solution design. We thought if we could, then we should, and so we did.

CHAPTER NINE

A Tale of Two Huaxis

From everything that I've written so far, you can probably already tell that I am not an anti-industry environmentalist. To my way of thinking, industry has been one of the key drivers in raising the overall human standard of living. Taking industry out of the equation or opposing it outright doesn't make much sense. Rather, I oppose *irresponsible* industry, and my opposition takes two main forms:

First, I think that industrialists who poison their neighborhoods should be forced to change their ways by whatever means necessary. Second, these industrialists should be made to clean up their earlier messes.

The adage that technological and economic growth requires a phase of dirty expansion has become outmoded. Modern technologies are cleaner and more cost-effective than they've been at any other point in history. Put differently, we now know how to manufacture things without poisoning our environment or our peers. We can afford to make products that don't harm our children or each other. We can and we should enjoy the fruits of modern industry. But while doing so, we should insist that our industries—all industries—operate under the safest, most sustainable, and most respectful paradigms possible.

For the most part, the best companies in the West do a good job of protecting their people, their consumers, and their reputation. As a result, they profit. They have learned to do this the hard way, weathering periods of experimentation, success, and failure, until regulations ensured their safest possible operational parameters. But these lessons are slow to take hold in the developing world, where governments fail to impose economic disincentives on bad behavior.

Which leads me to the Tale of Two Huaxis.

Most folks have never heard of Huaxi Town in Zhejiang Province, China, or the riot that took place there back in April 2005. I've heard some media outlets charge that Chinese authorities deliberately kept this incident under wraps. I consider it more likely that Westerners simply got their names mixed up. Because there's another prominent place in China that bears the same name—Huaxi Village in Jiangsu Province.

Far from being the site of anti-government riots, the Huaxi in Jiangsu has often been dubbed the richest village in China. And this may have caused some confusion, sort of like mistaking Moscow, Russia, for Moscow, Kansas. The first town boasts great architecture like Saint Basil's Cathedral, the Ivan the Great Bell Tower, and Terem Palace. The Moscow in Kansas has a couple of grain towers, the Antler's Bar & Grill, and a population of 314.

Just so I'm sure we're all on the same page, I'll use the term "Huaxi Town" when referring to the settlement in Zhejiang, and "Huaxi Village" when I mean the one in Jiangsu.

With that said, let's begin.

Up until a certain point, both Huaxis had much in common. Both, for instance, were tiny, and still are; Huaxi Town has a population of about 10,000, while Huaxi Village claims half that amount. Neither is what we might call a metropolis.

Up until the late 1970s, both Huaxis were backward boondocks in the doldrums of China's ancient rural culture. Oxen pulled carts along dirt roads. The surrounding wilderness touched up against hand-plowed fields and hand-me-down smallholdings. Tendrils of smoke rose from stone chimneys. Most citizens passed their lives without once leaving their immediate precinct. Then, in 1976, Mao Zedong passed away after more than forty years' reign as chairman of China's ruling Communist Party. At which point, many things changed.

Much has been written of life under Mao. I'll let other sources tell that story, since they do it much better than I can. Suffice it to say that a long period of backwards thinking with brutal consequences drew to a close. Mao's rival-in-waiting, Deng Xiaoping, seized control of the country and implemented sweeping reforms.

Perhaps the most famous of these reforms was a plan known as the Four Modernizations, or Mods, as most Chinese called them—national initiatives in agriculture, industry, national defense, and science. The Mods were certainly ambitious, possibly grandiose. Hardliners denounced them as being insufficiently Communist in their ideology. On the other hand, progressive thinkers celebrated them as "Leninist, with Chinese characteristics."

Regardless of which viewpoint you took, Deng's overall thrust seemed clear enough: to catapult China toward world prominence by the end of the twentieth century. To accomplish this, he was willing to tear down a lot of barriers, including—most prominently—the Bamboo Curtain, China's decades-old policy of secrecy and separation.

Through his Open Door Policy, Deng repaired diplomacy and trade with the global community. He legislated Special Economic Zones, or SEZs, in four coastal cities, each of which boasted unique manufacturing capabilities. Each SEZ offered tax privileges, reduced tariffs, and other incentives to foreign companies hoping to import capital goods and raw materials to Mainland China.

This arrangement became a smash success. Companies from

around the world rushed to fill the new market. Goods and commerce flowed into China. As a result, many Chinese saw their standard of living rise rapidly. Cities evolved at mind-boggling rates, along with the economies that fueled them. Industrial centers expanded, taking urban limits with them. The fingers of modernization began to reach deeper and deeper into China's ancestral hinterlands.

Eventually, these fingers touched both Huaxis, and altered them irrevocably.

Huaxi Village is close to Shanghai, a city justifiably famous for its status as an entrepôt. Ports like Shanghai generate trade, which in turn generates vast sums of money. Deng therefore allowed most of these cities and their outlying regions to conduct themselves almost however they pleased.

Led by their local Communist Party secretary, an innovative man, the residents of Huaxi Village established themselves as a sort of miniature corporation. Under the new arrangement, all citizens held shares in the town which, in turn, owned all the town's businesses, including a textile mill, a steel mill, agricultural facilities, and so on.

It took a few years for this new business model to gain traction. When it did, however, it paid off handsomely.

Case in point: In 1993, a destitute peasant, Ms. Ge Xiufang, relocated from northern Jiangsu. Her son had recently graduated university and, while searching for employment, he ran across a newspaper ad beckoning new workers to Huaxi Village. Within two years of arriving there, Ms. Ge, her husband, and son had become vested shareholders in the Huaxi Village community.

Their timing was perfect. Back then, the holding company that runs the town was branching out, investing in interests throughout the province. By 2011, the firm had grown to a $7 billion a year business that reportedly controlled fifty-seven subsidiaries and seven additional holding companies. In a style most Americans would envy, Ms. Ge and her family had gone from rags to riches in just under twenty years.

In 2011, the *New York Times* reported that Huaxi Village's 2,000 shareholders "live comfortably off their dividends." Ms. Ge was interviewed for that piece. She reported that she and her husband had gone from having nothing to living in a townhouse—accommodations most Chinese (many Westerners, too, for that matter) would find eye-popping. But that wasn't all. Their home was located a few doors down from their son's two-story house, which featured marble floors, a spacious aquarium, and other modern luxuries. The family owned three cars, Ms. Ge said.

Again, rags to riches.

The article went on to note that residents of Huaxi Village receive annual stipends, vacations, health care, and helicopter rides from a small fleet of airships the village had recently purchased, apparently just because it could.

By 2011, Huaxi Village had added tourism to its list of profitable businesses. Each year, some two million people arrived from all over China and indeed the world. Wandering the well-swept streets, they wondered at this economic phenomenon that had blossomed, quite incredibly, where Mao's brand of Communism had once choked off prosperity.

No doubt they also came to ogle the skyscraper that the village was building smack in the center of town. At seventy-four stories—taller than Manhattan's iconic Chrysler Building—the single tower was designed to feature a "a five-star hotel, a gold-leaf-embellished concert hall, an upscale shopping mall, and what is billed as Asia's largest revolving restaurant." Its total cost: about half a billion dollars.

I know what you're thinking. This looks very impressive. And doubtless, we should take Huaxi Village as the ultimate paradigm for socially invested industry. But if you think this sounds too good to be true, you may be right. Some detractors charge that Huaxi Village is quite possibly an outlier put forward and sponsored by the Chinese government. The Chinese Communist Party, these detractors claim, grants Huaxi unique subsidies that have facilitated its meteoric economic rise. After all, the holding company for Huaxi Village refuses to disclose its investments.

It's an interesting observation, but the most glaring criticism I've heard about Huaxi Village is one that indirectly involves the pollution agenda. While the vested residents of Huaxi Village work seven days a week (in the Chinese style), the real labor needed to sustain their utopian conglomerate comes from an army of workers ten times their number. Bused in from poorer outlying provinces, these workers do not share in the prosperity of Huaxi Village—the very prosperity they help to create. Nor does it seem that they will ever be granted that chance. There is no stock plan or profit sharing for the workers, though they reportedly shoulder the burden (as well as the dangers) of the many industrial processes that keep the vested villagers in clover.

As for the other Huaxi—that story is a little different. Up until recently, Huaxi Town, population 10,000, was a picturesque, idyllic hamlet of Dongyang City. Set near the center of Zhejiang Province, Huaxi (sometimes anglicized as Huashi) presented the sort of landscape painters dream about. Mountains rise east and north of the settlement while fields sprawl away from its western and southern flanks. A stream wanders down from the highlands to meander across the town's western edge. Bursting with natural beauty, the district once seemed like a kind of paradise, otherworldly in its splendor and revered by the people who lived there.

Until it became industrialized.

Zhejiang is one of mainland China's most affluent provinces, but it didn't develop like that overnight. In the early-to-mid-1990s, city officials began renting space on the western edge of town to thirteen large and profitable firms that manufactured chemicals.

These manufacturers erected plants to produce their products on what had once been Huaxi Town's prime farmland. By some accounts, the combined footprint of these plants spread across nearly 15,000 *mu*, or Chinese acres—about 2,500 common acres or 1,000 hectares, to use European units.

Soon after these industries arrived, and in a fashion sadly all too typical in many parts of China, they began to dispose of their effluents and solid waste improperly. The villagers of Huaxi Town watched, helpless, as their fruit trees refused to flower and then died. Crops they had planted nearest the factories wilted, then withered, then snapped from their stalks. Schoolchildren reported irritation in their ears, eyes, and throats. Many locals were rushed to hospitals with swollen limbs for reasons no one could ascertain at first. The stream that ran through Huaxi Town, which residents had once prized as something close to holy, the life spring of their ancestral home, turned poisonous. The number of stillborn babies spiked.

One villager later told the free press, "The chemical plants in Huaxi have seriously polluted the local environment."

Independent investigators conducted surveys. Many reported seeing drums bearing telltale skull-and-crossbones insignias stacked in caches near local water supplies. The few wastewater pools and disposal stations that had been built in Huaxi Town were constructed hastily, using inferior-grade concrete. Another inquiry revealed that local factory "specialists" were often little more than peasants hired from neighboring provinces and put to work without any training in even basic safety techniques.

"We have petitioned [our government] many times without getting any solution," another villager told a foreign journalist.

On October 16, 2004, an exposé called "Huaxi: What Caused You to Become a Disaster?" ran in the *China Chemical Industry* newspaper. The author, journalist Weng Goujian, detailed the rising scourge of pollution in Huaxi Town and its neighboring precincts. Though scathing and explicit, the article gained no traction. Following a well-known playbook, local officials simply ignored it. Their tacit dismissal of Weng's article and everything it stood for set the stage for all the unrest that followed in the spring of 2005.

Chinese citizens have increasingly mounted grassroots-style interventions when they've felt that their government was ignoring their pleas for justice. On March 23, some elderly residents of Huaxi Town took matters into their own hands by erecting a dozen makeshift bamboo awnings beside industrial service roads. The residents said they wanted to monitor the flow of traffic servicing the poisonous plants. They promised to, if necessary, block factory vehicles from transporting raw materials into the facility, and so-called "finished" materials out of it.

Their intervention began peacefully. Accounts say the elderly demonstrators stopped a few vehicles and asked what they were transporting, curious to know what kinds of chemicals were traveling through their community. Within twenty-four hours, however, local officials dispatched a mob of security guards, who cleared the roads by beating the elderly with sticks. Many observers noted that some victims were badly bloodied.

At this point, things took a turn for the worse.

Word spread quickly through Huaxi Town. Outraged by the manner in which their elderly had been treated, the residents flocked to the bamboo posts and manned them in greater numbers. Again, they blocked vehicles from entering and exiting the plants. This time, however, they seized a schoolhouse near the heart of town, which they used as their center of operations. Children pressed into service as messengers scurried this way and that on foot, delivering advisories, orders, and codes.

The situation remained static for about two weeks. Then, on April 10, municipal officials ordered a cadre of several hundred military personnel to open the roadways by force. Many security officers took this as a license to run over villagers using their government vehicles. According to Reuters, two women were killed immediately, but more deaths reportedly followed, including a slew of injuries.

In practically the blink of an eye, this once idyllic hamlet transformed into a full-blown war zone. Twenty thousand people from Huaxi Town and its surrounding precincts turned on the security forces, whose numbers hovered at 3,000. With few other weapons available to them, the villagers began to throw rocks. Scavenging weapons from fallen guards, they accumulated a deep supply of batons, steel helmets, rubber truncheons, and tear gas canisters. Security vehicles and public buses were also stolen, their windows smashed by bricks and stones; some were overturned to be used as barricades, or simply because it felt right for the occasion.

Town officials were ferreted out and beaten. Reports indicated that, as the tide of battle turned, security officers tore off their uniforms and attempted to blend with the crowd rather than face attack.

So what is the moral of our Two Huaxis? Both incidents exemplify industry run amok, and both hinge on the same basic principle: bad things happen when industries view themselves as separate from the environment they are housed in and the people who allow them to profit. You cannot ignore externalities forever. Industrialists who treat their communities with contempt are not only immoral but, in the long term, impractical. Stories like what happened (or could happen) in the Two Huaxis highlight the downside of industry.

By now, I hope you're asking yourself: What can we do about this? How we can level the playing field, and do so in ways that are simple and cost-efficient?

In the case of Huaxi Town, it seems clear that—like the leadership of so many other nations—the government of China needed to put its money where its mouth is. Specifically, they needed to stop working on environmental policy and get down to brass tacks: protect the environment. And they are.

In recent years, numerous examples have shown that the government of China is moving in the right direction. Sanitation programs are being installed across the country. Dirty industries are being shut down and relocated away from populous areas. Investments are being made in green technologies: buses powered by natural gas and scrubbers for the exhaust of coal-fired power plants. Do these actions amount to the tip of a distressingly large iceberg? Yes, but again, they show clear direction and, better than that, commitment.

Frankly, I don't worry so much over China. Their environmental situation is certainly bad right now, but in the long term, I feel like they'll figure it out. In the meantime, there is a lot we can do on more practical levels. For instance, good standards, good regulations, and good regulatory agencies need to be in place. For the sake of expediency, the country's immediate goal need not be 100 percent compliance. We have to realize that many industries will fall short of any mark set, especially in developing nations where bad actors routinely cut corners. No, the first step is to set the mark, and make it clear why the mark is there and how it will benefit all parties. Once that's done, compliance should follow over time.

Having a standard makes it glaringly obvious when parties fall short of it. At which point, government regulators step in to police their own industries. It is also critical that strong and credible NGOs take on the role of watchdogs where governments prove themselves unwilling or incapable of doing so.

So what other specific points should be part of this ideal paradigm?

First, industries should aggregate in areas detailed for that purpose alone. It is too risky to have manufacturing plants flung widely across a population. They should all be put in the same place: industrial parks, zoned for this purpose. This will minimize the footprint created by industrial processes, and therefore any negative impact that industries create on the overall community.

Second, these industrial parks should be set far apart from residential areas. Throughout my travels, I've noticed how the poorest citizens tend to end up in regions that are a) inexpensive to dwell in, and b) close to menial jobs. In many parts of the world, the spaces occupied by industry fit both these criteria. Ideally, a government-mandated buffer zone between industrial parks and the closest human habitat would satisfy this safety requirement.

Third, we must require the tenants of industrial parks to share resources wherever possible. Instituting and enforcing this policy serves a number of purposes, all of which benefit both the constituent industries and the public at large. Industrial park tenants should also provide support to the community by funding local schools and events, supporting health clinics, and the like.

A key component of the shared resource model should be the Common Effluent Treatment Plant. CETPs make especially good sense in cases where factories with similar production modes discharge the same general types of waste. The more specific the effluents, the more specific the filtering process required to detoxify them. Imagine a plant that filters all the exhaust and wastes, both liquid and solid, produced by an entire industrial campus. Essentially shared property, the CETP is paid for out of a till supplied by the tenant businesses' profits. With several factories footing the bill, the overall cost of creating and maintaining the facility decreases per party involved. The cost benefit alone is usually enough to appeal to industries.

Efficacy is another attractive feature of CETPs. Traditionally, CETPs run higher chances of producing discharges that are safe for public consumption. There's a simple reason for this: CETPs tend to use more up-to-date, real-time monitoring systems and treatment equipment than do independent operators. And obviously having an up-to-date system makes it far less likely that some nasty toxin will travel out of a discharge pipe to pollute the local river.

Ideally, a CETP maintains itself as a separate business. This streamlines and standardizes all systems while making it easier for government inspectors to come in and perform their audits. A shared CETP also incentivizes the sponsoring businesses to police one another. Why? Because the burden of violations discovered at a CETP, as well as any fines that accompany them, are shared by all constituents. When everyone has skin in the game, the overall trend heads north, toward compliance.

The shared resource model should also extend to all solid wastes that industrial parks produce. Solid wastes are chemical byproducts of industrial processes, many of which are hazardous to both human health and the environment. Such products should be sent to hazardous waste disposal facilities—sort of specialty landfills where the wastes get separated and stored according to type and toxicity. Handled properly, they pose little threat to the community, which of course is the goal.

At this point in history, most middle-income countries have established industrial parks for many sectors, along with corresponding CETPs and hazardous waste facilities. However, many of these parks don't function properly. Others don't correspond to the shared-resource model I've outlined. By this point in my career, I've traveled around the world and toured dozens of industrial estates. Most fall into a wide range of shades that exist between the ideal scenario and zero compliance. This is a shame, but the situation isn't hopeless—more like it's in a state of flux and moving (albeit slowly) in the direction we'd like it to go. We just have to keep on top of the situation and keep raising the bar a bit higher each year.

In 2006, Pure Earth tackled a cleanup project in the Indian state of Gujarat, in a village called Muthia on the eastern border of Ahmedabad City. Muthia is located next to the Naroda Gujarat Industrial Development Corporation, an industrial estate in Gujarat. Naroda has undergone a lot of changes in the last decade; India's current prime minister, Narendra Modi, worked hard to clean up industrial estates during his previous tenure as governor of Gujarat.

But there are still problems, and we found some of them in 2006. Industrial solid waste regulations were not working; vast amounts of toxic material had been dumped in the local river or on the shores of a local stream. By the time we arrived, approximately 60,000 tons of really nasty contaminated sludge had accumulated. I admit that it's not a passable technical term, but "really nasty" well describes the grotesque and gummy glacier of goo which had turned a worrisome shade of red-green, and which had spread over outlying areas due to repeated dousing by monsoon rains.

This sludge had been a decade in the making. Chemical analysis showed that the effluent had deposited a mix of heavy metals, solvents, and other compounds directly into the soil next to the river. These contaminants were leaking their toxic legacy into drinking water. A population of some 85,000 local people was definitely at risk.

With help from the local pollution control board, we were able to draw up a simple strategy for our pilot intervention: scoop up all the acutely toxic waste and send it to an abatement facility for immediate and proper disposal. Local industries were very helpful during this phase; they covered all heavy equipment costs and donated funds to cover the expenses incurred by the disposal facility. Once this was done, we tested the area again and were not surprised to note that toxins had permeated the soil and were therefore still present. Removing the worst of the waste had decreased toxin levels by about sixty percent, but the pollution that remained continued to pose a formidable hazard.

Enter the next phase of our cleanup, and one of our more interesting solutions, in my opinion. After conferring with specialists, Pure Earth funded several rounds of treating the soil with vermiculture. In other words: worms. We deposited thousands and thousands of earthworms at the site, but these weren't just any old worms. The particular breed we used grow to be about two feet long and have the unique capability of absorbing heavy metals and contaminants into their bodies.

So here's how it went. Once the nasty, heavily toxic goop was hauled away, plus a few layers of soil beneath it, we spread piles of straw on the ground and scattered these gigantic earthworms on top of that. We watered the straw to make it soft and to keep the worms from drying out before they began to burrow. And burrow they did.

The worms ate their way down through the soil and disappeared. For a couple of days, they did what earthworms do best, I suppose—eating and moving about, pushing through subterranean tunnels, making their way through the eyeless dark. But then they began to emerge again. Bloated and sickly, they wriggled back up to the surface, where they flopped over and lay on the turf, quite spent, and died. Their bodies had literally absorbed the soil's pollutants, which had killed them. All that was left to do was scoop up the straw, the dead worms along with it, and cart the whole mess to a hazardous waste facility to be disposed of as a contaminant. At which point, we repeated the process. More straw. More worms. More waiting.

After three or four such iterations, we tested the soil again and were pleased to find reductions of, on average, eighty percent in toxins from pre-intervention levels. But the best part was this: in all cases, the levels of pollution had fallen below danger thresholds. In fact, for some chemicals, the toxins had become undetectable throughout strata of soil, water, and plant. As of the writing of this book, that patch of ground in Gujarat, the same tract of once-contaminated soil, is used to grow crops safely that feed the local community.

This highly effective intervention cost Pure Earth $25,000 over two years. All things considered? I call this a tremendous win.

CHAPTER TEN

Dark Clouds on the Horizon

Out of the three main pollution pathways—air, water, and soil—air pollution is probably the worst. It comes in two forms: household air pollution and outdoor air pollution. Household air pollution is related to women and families exposed to smoke from cook stoves that use coal, wood, and dung; we'll cover this type of pollution in Chapter 11. For now, however, we'll concern ourselves with outdoor, or urban air pollution.

Outdoor air is pollution's poster child. We've all seen the images of smog-stricken Beijing, with visibility down to just 100 yards, and all the inhabitants wearing masks. Images of belching smokestacks representing uncontrolled industry have also burned themselves into our collective memories. Smoke and gray, hanging smog covering cities, causing everything from coughs to cancer—this is what we visualize when we conjure up thoughts of pollution.

The scientific community recently determined that, of the many toxicants found in polluted air, the deadliest are the smallest particles. Typically found in the gray smoke coughed from a poorly tuned truck or car, particulates less than 2.5 microns in size are inhaled deep into the lungs, all the way into the alveoli, the tiniest sacs that transfer oxygen to the blood. Once these particles enter the blood, they can cause all kinds of disease—cardiovascular illness, cancer, and more.

Don't get me wrong. There are other pollutants in outdoor air: ozone, sulfur, larger particles, and many specific chemicals. Each causes health problems, but PM2.5, the technical name for these small particles, is the class that causes the most concern. This is the stuff that kills kids.

Before I show you examples, bear with me through this brief lesson in relevant history.

Some 250 million years ago, sulfur-saturated magma burst through the earth's crust and overflowed the region we now call the Siberian Traps—the area around the northern tip of Siberia, just inside the Arctic Circle. The eruptions were a cataclysmic geological event and one that may have walked hand in hand with the so-called Great Dying that took place at the end of the Paleozoic era. Clouds of soot rose, blotting out the sun. The lack of light caused changes in the earth's climate and atmosphere—changes that, in turn, led to the extinction of seventy percent of our planet's terrestrial vertebrates and ninety-six percent of all marine species.

Now that's what I call air pollution. But of course, all this happened well before humans were around. So it's not relevant to our day and age.

Or is it?

The eruptions at the Siberian Traps continued for a million years, give or take. Scientists estimate that they produced more than a million cubic feet of lava, which cooled and hardened into vast plains of igneous rock.

Fast-forward several thousand years; the rock plains got covered by snow and ice.

Fast-forward some more, this time to the early 1920s. Russian settlers entered this landscape and, in the shadow of the mile-high Putorana Mountains, built a town on pilings driven through the permafrost. They named this town Norilsk.

They could hardly have picked a more difficult place to call home. Snow buries the countryside throughout most of the year. During the worst cold snaps, temperatures plummet to negative sixty-three degrees Fahrenheit. In winter, the sun doesn't rise for six weeks. But as harsh as these conditions are, mankind ended up making them even worse.

In the early 1930s, geologists surveying the region made an astonishing discovery. The ancient volcanic activity at the Siberian Traps had created the largest supply of nickel, palladium, and copper on earth. Norilsk stood right on top of it.

Eager to harness the mineral wealth, Soviet officials co-opted the town as a key node in their infamous Gulag prison system. When the labor camps opened for business in 1935, Norilsk housed some 1,200 inmates. Two years later, when Stalin began his Great Purge, this number jumped by a factor of seven; at this point, it became painfully clear that most of the prisoners in the Gulag had been exiled there for political reasons. Occupancy continued to rise until 1951, when Norilsk reached peak occupancy with more than 72,000 prisoners. According to some estimates, half a million prisoners were forced to work there between 1935 and 1956.

These prisoners were all exploited as slave laborers to build facilities for (what is today) the Norilsk mining-metallurgic consortium. The largest heavy metal smelting complex in the world, Norilsk has also become one of the most polluted cities in the world.

These days, anyone approaching Norilsk will likely notice the dead zone first. Acid rain falls in an area roughly the size of Germany, with Norilsk serving as the area's epicenter. All vegetation within thirty miles of the city has withered. The air is so thick with particulates and sulfur dioxide that almost nothing will grow.

Closer to the city, the snow turns a number of foreboding shades: pink and yellow, then sooty, then gray, and finally black as tar. By the time you reach this point, it's easy to see what's causing the damage. A line of oblong factories rides the horizon like giant dominoes tipped over onto their sides. Distant smokestacks belch dense columns of grime that blot out the sun.

If you look up Norilsk on Google Maps, it appears as though the satellite image is obstructed by clouds. It's not. Those "clouds" are the smoke from Norilsk's production facilities, ever belching out pollution. No satellite can secure a clear shot through that haze, no matter how many passes it makes.

Each year, these smokestacks release over four million tons of cadmium, copper, nickel, arsenic, selenium, zinc, and lead into the atmosphere. Some researchers estimate that fully one percent of sulfur dioxide emissions the world over originates in Norilsk. By this point in history, smoke from the consortium has deposited such massive quantities of heavy metals in the surrounding landscape that, in recent years, mining companies have turned profits from processing the soil around the city.

The population of Norilsk hovers at about 170,000, almost all of whom work for the mining consortium. What choice do they have? The city's extreme isolation has made it an economic closed loop. Property values are pitiful since nobody wants to move there. By and large, the residents can't make enough money to leave.

People live short, hard lives in Norilsk. Extreme pollution takes its toll. Life expectancies for city inhabitants are ten years less than the Russian national average. Almost everyone suffers from respiratory ailments, blood and skin disorders, cancer, you name it. But among the most shocking statistics you'll read about Norilsk is this: only four percent of the city's adults are considered clinically healthy.

If you think this is a remote example of the dangers posed by air pollution in Russia, there are many others. Magnitogorsk, for instance, with its steel smelters polluting a population of 80,000. And there is Cherepovets, a remote town on the Volga

River some 300 kilometers northeast of Moscow. The name Cherepovets means "city of skulls," a term derived from ancient times when a skull-like hill played host to the temple of Veles, the Slavic God of Death.

In modern times, the name "city of skulls" holds unintended significance. Cherepovets is home to the Severstal steel plant, the largest iron smelter in Russia, whose smokestacks earned the sobriquet "Chimneys of Hell." Picture these massive flues; they hover over the city like the towers of black magicians, pumping out toxic smoke that blankets the town.

In 1965, in a tacit admission of the Severstal plant's toxicity, Russia's Ministry of Black Metallurgy established a "sanitary secure zone" 5,000 kilometers in diameter around the facility. Soviet authorities promised to relocate people living within this zone, but they never did. Thirty years later, Nadezhda Fadeyeva, a fifty-six-year-old mother of three, still found her name on the waiting list. Finally, she had had enough. She took her case to the international courts.

In a 2006 lawsuit filed with the European Court of Human Rights, Fadeyeva claimed that the Severstal plant accounted for ninety-five to ninety-seven percent of Cherepovets's industrial emissions, and also that these emissions were way off the charts. According to one study cited, exposure to pollutants had caused residents of Cherepovets to experience increased incidents of headache, thyroid abnormalities, cancers of the nose and respiratory tract, and chronic irritation of the eyes, nose, and throat, as well as impacts to neurological, cardiovascular, and reproductive systems.

Russian authorities eventually settled Fadeyeva's case. That alone is worth celebrating, since the outcome implies some measure of assumed responsibility. Today, in fact, legal experts cite Nadezhda Fadeyeva's lawsuit as "[having] the potential to fundamentally shift thinking in Europe on the connection between human rights and the environment."

But I don't believe it necessarily has. Because, sadly, air pollution has become the norm for modern industrialized countries. Mining and manufacturing businesses certainly cause a lot of the damage, but transportation also contributes, especially where vehicles burn cheap fuel in poorly maintained engines. Exhaust from these modes of combustion creates remarkably carcinogenic particles that spread like spores through the atmosphere and can damage whatever organic tissues they touch.

◆

China presents a well-known example of this situation. The People's Republic has become infamous for its air pollution woes. Its storied and rapid industrialization created a multiplied backlash effect that is most clearly seen in the purity of its air.

Harbin is China's tenth most populous city with a population of eleven million. To put this into perspective, the number of Harbin residents nearly equals the number of residents in New York City and Los Angeles combined. But in late October 2013, the entire city was forced to shut down when visibility decreased to less than fifty meters due to rolling banks of extraordinarily dense smog. As reported by the Associated Press, readings taken during this crisis showed air pollution spiking to levels forty times greater than maximums set by international safety standards. For air pollution, that's an astonishing feat.

But was anyone really surprised by this? China has fast become the world's number-one producer of coal. Coal accounts for seventy percent of the nation's overall energy consumption. In July 2013, research published by the Proceedings of the National Academy of Sciences stated that Chinese life expectancies have dropped on average by 5.5 years since the proliferation of coal. The point seems clear: you can't burn that much coal without paying the piper down the line.

It's worth noting that China's problem isn't localized; it's spread all over the world, though it shows up most often in Asia and the Far East. For instance, at the writing of this chapter, Ulan Bator, the capital city of Mongolia, claims the dubious title of having the worst air pollution in the world. Mind you, Ulan Bator isn't very large. It has a population of 1.3 million people and, interestingly, an elevation of 4,300 feet, or nearly the same as Grand Junction, Colorado. It's also way out in the middle of nowhere; the nearest major city is probably Beijing, some 700 miles to the southeast.

How can such a small, elevated, isolated city have the worst air pollution in the world? Because of its primary heating source: inexpensive, dirty coal.

Beijing doesn't fare much better, of course. In fact, in November 2010, China suffered particular indignity from a now-infamous message posted on the Twitter account of the U.S. embassy in Beijing. Foreign policy experts say the embassy has maintained an air-quality monitoring device on its roof for years. The device records its measurements using a U.S. EPA scale in which a score of one indicates the lowest possible air pollution and 500 indicates the maximum. Levels between 300 and 500 are considered "hazardous."

On that infamous day in November 2010, the device offered readings over 500.

"Crazy Bad," tweeted @BeijingAir. Soon after that, the post was removed suddenly and without any explanation. If only air pollution problems worldwide could be deleted so easily.

In January 2013, as reported by the *New York Times*, the same meter on the roof of the U.S. embassy hit 755 on a day "when all of Beijing looked like an airport smokers' lounge." Put differently, the quality of air in China's capital city went from really bad to much worse in a little over two years.

Despite all this, there are definite signs that progress is being made in China. In February 2014, the UK's daily newspaper, *The Guardian*, showed that Chinese scientists had become more outspoken about their country's environmental conditions. Breaking with long-held, traditional silence, these scientists had begun to offer dire warnings of their own.

The Guardian's article cited He Dongxian, an associate professor from the College of Water Resources and Civil Engineering at China Agricultural University. Dr. He put forth research that she summarized like this: if left unchecked, China's smog would afflict the nation's agricultural cycles "somewhat similar [in manner] to a nuclear winter."

Specifically, Dr. He's studies showed that air pollutants block the amount of light crops receive by about fifty percent. This lack of light impedes photosynthesis, making it difficult for plants to grow. Dr. He compared seeds grown with artificial light in a laboratory to seeds of the same variety grown in a greenhouse located in suburban Beijing. The lab-grown seeds sprouted in just under three weeks, while the greenhouse seeds took over two months—more than three times the lab seeds' duration.

Dr. He also noted that the greenhouse seeds appeared sickly. "They will be lucky to live at all," she said.

It's worth noting that bad air comes and goes. In fact, from what I've seen while traveling in China, the most ferocious bouts of really terrible air occur during periods when atmospheric conditions conspire to lock toxic fumes in one place. This, of course, sets a dangerous precedent. It's never a good thing when common meteorological activity—a lack of wind or a common thermal inversion, for instance—can bring a large city to its knees.

Consider Beijing during the 2012 Olympics. Back then, the city got almost unreasonably lucky. Two weeks before the games were scheduled to start, the air in Beijing was so thick that pedestrians couldn't see two city blocks ahead of themselves. I was there at the time and worried sick over the athletes. How could they represent their countries in sport when it hurt just to draw shallow breaths? What damage might they do to their finely trained bodies by subjecting them to such unhealthy conditions? But right before the first contests began, a storm blew through, the murk parted, and blue skies appeared above the city. Just in time, coincidentally, for the cameras of international news teams to start rolling.

Despite the clear air in Beijing that day, regardless of what the cameras filmed, I didn't see many Chinese nationals removing their breath masks to enjoy the weather. Nor was China's government willing to acknowledge that the air during the Olympics was not the norm.

But as I say, good fortune aside, changes are starting to happen. China's stance on air pollution took a recent and marked turn in the 2010s. The People's Republic launched an aggressive program to combat its outdoor air pollution. The program features dozens of separate initiatives geared at reclaiming breathable air.

Since about seventy percent of China's air pollution comes from tailpipes, a big push has been made to switch automobiles to cleaner-burning fuel. Municipalities all across China have converted their bus fleets to run on compressed natural gas. Strict emissions standards have been created and enforced with rigor nationwide. Larger industries once based in and around Beijing have been moved to less-populated outlying regions. Factories have been issued controls for their stack emissions; violation of these controls invokes the heaviest penalties the Chinese have ever considered levying.

China has a long way to go, especially since, per capita, air pollution worldwide and the number of fatalities stemming from it are growing fastest in Asia. At this point, fully sixty-five percent of deaths in Asia arise from complications related to air pollution. Each year, polluted air claims the lives of three-quarters of a million people in China alone. These are staggering statistics.

But most other countries have barely begun to address the problem. Air quality is terrible in Manila, Ulan Bator, Dhaka, and many cities in Bangladesh, Iran, Pakistan, and Afghanistan. African cities are not far behind. So, while China is making inroads, the rest of the world is getting worse.

At the time of this book's composition, India serves as a prime example of a country in need of a paradigm shift. Air pollution in many Indian cities has become as bad as that of their Chinese counterparts. At present, some twenty-five percent of India's deaths per year arise from complications stemming from air pollution. A *New York Times* article reported that, during the first three weeks of January 2014, air meters in Delhi averaged scores of 473 on the EPA scale. A score of 473 might not sound like much after I mentioned Beijing's 755. Keep in mind, however, that Beijing's whopper was an isolated incident, while Delhi's January 2014 ordeal stretched out for three whole weeks. And that score of 473 was twice the average reading in Beijing.

China and India are the two new global superpowers-to-be. They need to compete to maintain healthy relationships, but surely not over who has the most toxic air. That's a contest I think we can all do without.

Here again: credit where credit is due. I should note that many cities with truly filthy air—Delhi, Manila, and Ulan Bator—have publicly recognized their air pollution problems. But talk only goes so far. It's nice that some plans exist here and there on paper. It's encouraging to think that these same plans might mitigate the problem of air pollution over time. But when will they be implemented? When will the numbers improve?

While governments and politicians dicker, the sick and the elderly die in droves. I've seen the effects of these conditions personally: nearly four million people dead each year from outdoor air pollution. These numbers won't get much better, even if we act now. Already, scientists expect at least six million people per year to perish from outdoor air pollution by the mid-century mark.

Quite likely, as you read this, you consider yourself immune to this problem. After all, quite likely, you live in the West. The problem with that sort of thinking is that conditions here in the West aren't all honey and roses either.

In sunny California, the state board of education calls asthma the leading cause of student absenteeism. According to many researchers, this asthma has roots in air toxicity. When our air becomes choked with microscopic particulates and gases—watch out. Especially if you're a child. Like all the other types of pollution we've covered and will cover, air pollution sinks its ugly hooks into all human beings exposed to it. But the small, developing bodies of children prove especially vulnerable to its onslaught.

Also in California, some 5,000 premature deaths occur each year that are attributable to diesel exhaust alone. This number represents a fifth of the 25,000 deaths a year that California experiences due to air pollution in general. These 25,000 deaths strain the Golden State's medical infrastructure by an estimated $200 million a year. And when you consider air pollution among all fifty states, California leads the pack with about half the number of air pollution-related deaths in America each year.

To date, Pure Earth has played a limited role in combating outdoor air pollution. Since inception, our focus has stayed on contaminated soils and sites. Still, we have had a few successes. One of our earliest projects in this sphere began back in 2001, in Senegal. At the time, the country was still using leaded gasoline, emissions from which had polluted their air.

The approach we took was simple: we worked with the Senegalese government, the United Nations Environment Programme, and other organizations to initiate a phase-out of lead in gasoline. Simultaneously, we initiated a system by which we could measure the lead levels present in gasoline being pumped at gas stations across the country. This second phase gave us direct insight into the efficacy of regulations banning the import of lead fuel, which our team had drafted.

It took two and a half years and a lot of meetings, but eventually Senegal could tell the rest of the world it was fully free of leaded gasoline. In fact, our intervention worked so well that we helped a few other countries—Mozambique and Tanzania—obtain similarly good results. Nowadays, leaded fuel is almost completely gone from the world. Only North Korea and Myanmar still use it, according to latest reports.

If only every air pollution problem could be handled so easily. When air pollution reaches pandemic proportions, as it has in the cases of Russia, China, and India, results are not so easily achieved. But we have to keep working. We have no choice, since the stakes apply to us all.

CHAPTER ELEVEN

The Silent Killer in Household Kitchens

Most people hear the term "air pollution" and immediately think of things we discussed in the last chapter: crowded urban landscapes choked by vehicular exhaust, smog settling lazily over the Los Angeles basin, or the smear of black gunk that we find on a tissue we've blown our nose into after a brisk walk through Mexico City on a smoggy day.

Images of decrepit manufacturing facilities pop into our brains. Air pollution means power plants gushing soot from their smokestacks, coal-fired grease clouds blotting the sun. Pollution has fouled our air since the dawn of the Industrial Age. By now, we've practically come to expect it, sad as that is to say.

But everything I've just mentioned exemplifies *outdoor* air pollution. What most people don't realize is that polluted *indoor* air—and the air inside a home, in particular—is frequently more hazardous to human health.

This may come as news if you live in the West. Here we have governmental agencies that devise and enforce environmental standards: the U.S. EPA and its local counterparts, as well as the Occupational Safety and Health Administration. We take it for granted that Western-style buildings have smoke and carbon monoxide alarms. Doesn't every building on earth feature sophisticated heating, ventilation, and air-conditioning systems equipped with high-efficiency particulate air filters? Isn't every kitchen on earth constructed to vent stove and oven exhaust properly?

The answer: absolutely not.

Let's suppose you're one of the billions of people who live in a developing nation. If you are poor, more than likely you heat your home, boil your water, or cook your meals over an open fire. Perhaps instead, you employ some sort of simple, traditional stove. It doesn't matter. By either method, you're probably subjecting yourself and your family to smoke inhalation on a routine basis. Open fires and simple stoves are renowned for poor ventilation.

This would be dangerous in and of itself. To complicate matters, however, the fuel that you burn is remarkably crude, commensurate with your economic status. In 2000, a bulletin on indoor air pollution produced by the World Health Organization stated:

> In general, the types of fuel used [by people in developing nations] become cleaner and more convenient, efficient and costly as people move up the energy ladder. Animal dung, on the lowest rung of this ladder, is succeeded by crop residues, wood, charcoal, kerosene, gas and electricity. People tend to move up the ladder as socioeconomic conditions improve.

People burn animal dung? Indeed. The sun-dried leavings of a family's subsistence herd often fuel their cook fires. Crop residues, the next most glamorous category of fuel, amounts to stalks, cobs, and husks left over from plants farmed for subsistence and commerce: corn, bamboo, switch grass, eucalyptus, and so on.

The WHO report leaves out that, in cases of extreme poverty, certain populations in the developing world have also burned their own garbage for fuel. By now, however, I'm sure you're getting the point: burning such fuels can be dicey at best, while improper burning of such fuels makes a bad situation worse.

Smoke produced by any of these agents can generate fine particulate matter over 100 times the maximum safety levels established by experts in human health. When corralled indoors, this smoke becomes especially toxic. Exposure to it over time can heighten the risk of pulmonary disease, respiratory infections, and other afflictions.

Women, the elderly, and very small children fall prey to these conditions most often, since men and boys are frequently gone from the household, pursuing work. Some studies show that women in rural areas of some developing nations spend upwards of seven hours a day within two meters' proximity to their household's poorly ventilated cook fire. As such, they suffer exposure to pollutants contained in the fire's exhaust. Chief among them: particulates, benzene, carbon monoxide, potassium (which in abundance can be lethal), and methyl chloride.

During cold weather, those seven hours of exposure can expand to encompass "a substantial portion of each twenty-four-hour period," according to the WHO report.

Of the three demographics most commonly exposed to indoor pollution—women, the elderly, and children—children bear the brunt of disease. This should come as no surprise, given the territory I've covered already. A child's tiny, developing body simply cannot withstand such adverse conditions. The unique physiology of children makes them ingest pollutants far more readily than adults do. Their bodies also retain ingested pollutants for much longer periods, and with consequences far more dire than those experienced by adults.

In fact, childhood respiratory infections have become one of the leading causes of death in children under the age of five in developing nations. Beyond that, however, household air pollution has also been linked to cases of lung, nasal, and throat cancer, cataracts, tuberculosis, asthma, stroke, chronic obstructive pulmonary disorder, low birth weight, and infant mortality.

In 2000, WHO stated that household air pollution causes some two million excess deaths in developing nations each year. In 2012, however, WHO updated that number to about four million. (It also specified affected locales: statistically, populations in the Western Pacific and Southeast Asian regions suffered from household air pollution the most. The 2012 bulletin also noted that further research had strengthened the links between polluted indoor air and the host of ailments I've already listed.)

Let me repeat that pivotal finding: two million deaths rose to *four million* in a little over a decade. This kind of trend is both clear and upsetting. No wonder household air pollution is one of the single largest environmental health risks on the planet. A white paper produced in 2012 by the University of Pennsylvania agrees:

> . . . Significant financial resources from the global health community have been mobilized to tackle developing world health problems, almost doubling from $11 billion in 2000 to $22 billion in 2007. A large share of this assistance has gone toward preventing and treating specific diseases. Approximately a third of development assistance for health goes toward HIV/AIDs, Malaria and Tuberculosis (IHME 2009). In low income countries, these three disease groups together accounted for 1.6 million deaths, comparable to the two million deaths attributed to indoor air pollution health risks in 2008.

The numbers seem clear to me. Household pollution kills more people a year than HIV/AIDS, malaria, and tuberculosis combined. Yet very little has been done to address the problem.

Complex interrelationships exist between indoor air pollution in the developing world and other forms of pollution or environmental degradation. In Kenya, for instance, coal or biomass fuels (dung, wood, or vegetable matter) are burned on a regular basis. Smoke trapped indoors creates conditions where women and children suffer diseases. It also, however, increases outdoor air pollution since, as one might expect, air quality in a region tends to get worse overall when large populations consistently burn crude fuels without adequate filtration.

This increase in outdoor air pollution carries its own set of diseases and pathologies. But the problem doesn't stop there—not yet. Because while such countries are poisoning their air, they're also depleting the natural resources upon which their overall ecosystem depends.

Staying with the example of Kenya, a 2005 study released by the Global Village Energy Partnership cited that about thirty-eight percent of households in "high agro-ecological zones" burned animal dung and agricultural byproducts. Locals couldn't lay hands on actual wood, which had become a scarce commodity. Like the rainforests of Brazil, the Kenyan savanna has suffered attacks in recent years. And just as in the case of Brazil, we can blame the situation on unregulated agricultural expansion and land conversion. But the large-scale switch from wood to other forms of biomass creates a linked set of additional problems.

Dung and agricultural byproducts act as potent, natural fertilizers. Leaving them on the ground allows them to decompose and feed the soil from which a nation's crops arise. By gathering and burning these items, the Kenyans disrupt the natural cycle of growth and renewal. Crop yields fall, grazing lands wither, and the "cumulative effects on national food and security" (to quote the GVEP report) turn negative.

Kenya isn't the only problem zone, of course. Similar conditions exist in Nepal, where firewood reportedly accounts for seventy-two percent of the population's total energy consumption. Again, the chief problem is the traditional stoves most Nepalese use to boil water and cook their meals.

The online edition of IRIN in 2008 discussed Ms. Sitadevi Khadka, age sixty, a resident of Basamari, a rural village 200 kilometers west of Kathmandu. Ms. Khadka said she spends several hours a day cooking meals for her family over the wood-burning stove in her home. In Basamari, gas or kerosene-burning stoves are considered luxury items, completely unaffordable. As a result, Ms. Khadka had scavenged firewood from the local forests for years and resigned herself to a life spent coughing and rubbing her eyes in a kitchen choked by thick gray smoke.

"I know it's not healthy with so much smoke inside our kitchen," she told IRIN. "But there is no other option."

The most recent reports released by WHO suggest that some 8,700 Nepalese die prematurely each year from diseases related to household air pollution. This figure represents more than three percent of the overall number of deaths, a quantity that beckons attention.

"Nepal's household pollution is much worse than its outdoor pollution," said Gopal Joshi, program coordinator for an NGO called Clean Energy Nepal. "So one can only imagine the dangerous impact on the household members' health."

Doubtless, statements like this are the reason that Nepal's prime minister, Baburam Bhattarai, recently pledged that his country would have smoke-free dwellings by 2017.

At the risk of restating the obvious, death and disease from indoor air pollution aren't problems that affluent people face. Reliance on low-grade fuels and unclean forms of combustion are hallmarks of poverty, and Sitadevi Khadka is no exception to this rule. Along with some ten million of her countrymen—about thirty-eight percent of the overall population—she and her family live on the equivalent of less than $1 American per day. Extrapolate the number of people who share their socioeconomic station worldwide, and you start to see how billions of lives are suddenly imperiled by the simple act of attempting to make breakfast.

Okay. That's the bad news. Here's the good: solutions that vastly improve people's lives can be implemented rather quickly and for markedly modest sums. In fact, in most of the cases I've seen or heard about, technology isn't the barrier to implementation as much as the sociological and economic factors currently at play in each region.

Pure Earth itself has not been working in this area of concern—we've been active in the issue of polluted soils and waters. But there are wonderful NGOs that have been making a difference around the world, working with local communities and governments to save lives. I'll give you some examples.

Practical Action is a UK-based international NGO that leverages technology to alleviate problems caused by extreme poverty in developing nations. Prior to 2003, the NGO studied Wau Nour, a struggling district of Kassala in the country of Sudan. Its research eventually suggested that eighty percent consumption of biomass fuels was the norm in Sudan.

The following was found in a study the group released on this topic:

> Prolonged exposure to biomass smoke is a significant cause of health problems, including acute respiratory infections (ARI) in children, chronic obstructive lung diseases such as asthma and chronic bronchitis, lung cancer and pregnancy-related problems. The prevalence of these diseases in Sudan indicates the need for changes in practices and attitudes with relation to biomass fuels.

In pursuit of this "need for changes," Practical Action developed a three-pronged approach for their intervention in Wau Nour.

Their first point of attack was technological. The NGO tested pre-intervention domiciles and found that each registered high levels of particulate matter and carbon monoxide—a very bad sign. The households, therefore, had to be cleaned.

Smoke-blackened walls and ceilings were scrubbed to get rid of exhaust buildup, which, in many cases, proved extensive.

The design of each dwelling also fell under scrutiny. How could the position of stoves, for instance, be altered to facilitate ventilation? What kinds of chimneys or exhaust hoods could be used to vent cooking smoke properly? Once these factors were addressed, each house in the pilot program was equipped with the most important technological intervention of all: a simple stove outfitted to burn liquefied propane gas.

Why LPG? For one thing, it has a proven track record. LPG was first produced in 1910 and hit commercial markets two years later. It's been around for over 100 years and, in all that time, it's acquitted itself as a trustworthy, adaptable, portable, and proven solution to hundreds of mass market uses. For this and other reasons, car companies have begun to experiment with LPG as fuel for automobiles.

It turns out that LPG made a good fit for the Sudan. Local factories produced the fuel in quantity, so supply would never be an issue. And LPG burns cleanly—remarkably clean when you compare it to biomass combustion. LPG also generates more heat than biomass does, and does so more quickly.

With the technology arm of its intervention settled, Practical Action moved on to its second thrust: attacking the socio-economic hurdles of switching to LPG in Wau Nour. This may sound simple, but it's not; it's also the most important step in the process. Scavenged wood can be burned for free, but LPG costs money. Convincing an underserved population that they should pay for something they never had to before can quickly become a maze of logic.

Practical Action was smart, however. First it set up local microcredit facilities to defray any upfront expenses incurred by families switching to the new stoves or purchasing fuel. Then it aimed to set new policies that would benefit LPG stove owners moving forward. The icing on the cake arrived when the

Sudanese government offered to subsidize purchases of LPG fuel by fifty percent. This kind of incentive certainly encouraged people to adopt the technology. The government also exempted appliances that burned LPG from import tax. With these and other factors in place, a lifestyle that included cooking and heating with LPG suddenly drew within reach of people who hadn't been able to afford it previously.

But Practical Action didn't stop there. Its third and final thrust at the problem aimed at educating the public. Pure Earth also adopts this practice in practically every project we initiate, and I can't begin to tell you how influential this stage can be.

Change takes place very quickly when you sit down with the women of a community, many of whom are mothers of small children, and show them how something they're doing is hurting their kids. Contrary to what you might assume, there's hardly any shame involved. The women have lived with these conditions for some time; they already know that what you're telling them is true. They've seen their kids coughing and scratching their eyes. In many cases, they've lost a child. Their primary goal is to keep those still alive as healthy as possible. They're willing to hear you out.

As part of its information campaign, Practical Action set up training programs designed to demystify the health risks associated with indoor pollution, explain how best to ventilate smoke from a dwelling, and tutor owners of new stoves in how to use their appliances safely.

Was the program successful? Indeed, it was.

Using the method described above, Practical Action addressed indoor pollution in 167 households. These households experienced such an overall reduction in harmful smoke that another 137 households requested similar assistance. Which, in turn, prompted the municipal government of Kassala to embrace LPG stoves as the best practice for heating equipment in household kitchens.

Wau Nour then applied for and received clearance to create its own LPG refilling shop. And while all this was going on, public information campaigns and word of mouth were creating a snowball effect. Very quickly, the old method of cooking by traditional methods became unfashionable, regarded as backwards. The practice of cooking with LPG stoves became the new norm in Kassala.

There are other examples with good successes, all following a similar path. In Ghana, the burning of solid biomass fuels is off the charts. Estimates provided by WHO said that ninety-five percent of Ghana's population burned biomass, a figure that is higher than the working estimate of 73.5 percent for the northwest portion of Africa.

Typical rural Ghanaian families live in crude one- or two-room homes that are almost always poorly ventilated. Cooking is done either inside or outside the dwelling. Three-stone fires are often employed, an ancient arrangement so named because three stones of approximately the same height encircle a central blaze. The stones reflect heat back toward the center in an attempt to maximize thermal efficiency. Pots can be balanced on top of the stones and solid fuels fed to the fire through gaps between them.

Apart from being notoriously unsafe (users often get burned or scalded while working), a three-stone fire is also one of the worst producers of smoke. I should note that, in Ghana, some families had begun working with rudimentary clay ovens, but here again the amount of smoke produced was untenable, as was the fact that the smoke almost always passed freely into the family's abode, or the dwellings of other families.

The same environmental devastation that was described for Kenya could be found in Ghana. Wood scavenging had contributed to soil erosion and deforestation of the savanna. Burning sun-dried animal dung had diminished soil fertility. The Ghanaian ecosystem had begun to drift out of balance. It could not return to stasis unless an intervention occurred. But what could be done?

A project was undertaken by Enterprise Works Ghana (a worldwide NGO dedicated to fighting poverty in Africa), Shell Foundation, and the United States Agency for International Development. The partnership provided units called Gyapa stoves to people living in and around Ghana's two primary cities, Accra and Kumasi.

Gyapa is the Ghanaian word for "good fire." Gyapa stoves live up to this name by burning charcoal within a ceramic-lined fuel tray. The overall design harks back to the so-called "rocket" stoves first theorized by Dr. Larry Winiarski of the Aprovecho Sustainability Education Center in the mid-1980s. Compact, simple, and easy to operate, stoves of this type are renowned for producing remarkably efficient combustion as well as high temperatures from minimal fuel.

The term "rocket stove" has since evolved into a generic term that can be applied to several unit designs. At least one of these designs won Ashden Awards for sustainability practice in 2005 and 2006. Rocket stoves have been employed to good effect for cooking and heating water in Rwandan refugee camps and at sites requiring energy-based interventions in Zambia, Mozambique, Uganda, Malawi, and Zambia.

The Gyapa stoves proved successful in Ghana for a number of reasons. In the first place, burning charcoal is a more sustainable practice than scavenging biomass. Also, the Gyapa unit's ceramic fuel tray maximized fuel efficiency while other features on the apparatus helped to minimize and direct the exhaust from combustion.

The Gyapa stoves were certainly innovative. Of all the cook fire replacement stoves available at the time, they were among the best. But technology keeps evolving, as of course it should. Several designs for new units leave the Gyapa stove far behind. What will be harder to repeat in future projects, however, is the focus on engendering a self-sustaining socio-economic engine.

The team wasn't just equipping homes with Gyapa stoves. The project created an entrepreneurial, market-driven model that supported the profitable manufacture, distribution, and use of a healthier and more efficient technological alternative to cook fires (say that five times fast, I dare you!). This model put people to work at good-paying jobs on one end while cleaning up indoor pollution on the other. It was a win-win situation.

Here's how it worked.

Enterprise Works Ghana seeded the pilot communities by providing local craftspeople and entrepreneurs with training, plans, and funds to build their own Gyapa stoves. Once metal-workers and ceramicists figured out how to work together and producc Gyapas profitably, they distributed the units to retailers, who sold them to families that were eager to have them.

Why were Ghanaian families suddenly eager to have Gyapa stoves? Because their interest was piqued by a concerted public information campaign. Education is often the key to so many of the world's problems. Here again, the old saw holds true: Give a man a fish and you'll feed him for a day. Teach a man to fish and you'll feed him for a lifetime. Once the team showed local peoples how their traditional cooking techniques caused so many adverse health effects, they flocked to embrace new solutions. The partnership made certain that each family that purchased a Gyapa was given instructions for its proper use, and that customer care, quality control, and general business advice was given to the manufacturers and retailers.

So what was the upshot?

In 2008, its first full year of operation, the new business created jobs for 400 individuals and sold 68,000 Gyapa stoves. Sales for the following year were projected to nearly double. Meanwhile, the project team retested the air in homes that implemented Gyapa stoves. The results showed that air quality had improved by forty to forty-five percent. (Later tests conducted by independent agents boosted this figure to fifty-two percent.) With figures like these, they could reasonably state that children under five years old were twenty-five percent less likely to die from respiratory diseases.

This is a very good start. But unfortunately, concentrations of particulate matter in participatory homes remained well above WHO guidelines. Which means there is more work to do. New innovations are being considered to help secure better compliance. One of these innovations that I personally find compelling is the use of solar-powered cookers.

In 2007, Catlin Powers was studying climate change in the Himalayas of Qinghai Province, in rural western China. There she met a family who posed an interesting question.

"Why are scientists so concerned with outdoor pollution?" they asked. "The pollution in our homes is much worse. Don't you agree?"

A PhD candidate at the Harvard School of Public Health, Powers ran some tests on the air in the family's home and found it ten times more polluted than the air in Beijing. Clearly, something had to be done. But what?

A little while later, while still in Qinghai, Powers met another American. An electrical engineer and computer scientist trained at M.I.T., Scot Frank had come to China hoping to teach engineering in Beijing. When no students enrolled in his course, he had drifted to Qinghai. The conditions he saw there dovetailed with those that Powers had witnessed. Together they founded a company called One Earth Designs. The chief product this company sells is a parabolic solar cooker: a simple, portable, fuel- and emissions-free cooking appliance called the SolSource.

A SolSource unit looks like a small satellite dish lined with five mirrored panels curved like the petals of a flower. The dish sits on a stand equipped with joints to facilitate easy directional adjustments. An arm extends across the dish's concavity. At the center of this arm is a heat- and rust-resistant circular stand where someone can comfortably place a pot or frying skillet.

By now you can probably guess how this works.

Point SolSource toward the sun, and its mirrors will do the rest. The polished panes bounce light toward the center of the dish, directly at the pot stand. According to the company website: "On a clear sunny day, SolSource boils one liter of water in ten minutes, reaches grilling and baking temperatures in five minutes and searing temperatures in ten minutes."

Baking, you ask? Absolutely. You can bake by riffing on the concept of a Dutch oven. Place a baking tin inside a large black pot, place the large black pot on the pot stand, and voilà. Your biscuits, cornbread, brownies, or cake will be ready in no time flat.

The SolSource cooker can be used safely in a variety of conditions. It's even useful on cloudy days. According to the One Earth Designs website: "If you can see shadows, you can cook. If clouds frequently pass by, it will make cooking more difficult."

And even though they reflect light with eighty-eight percent efficiency, the mirrored panels remain cool to the touch. The company recommends, however, that SolSource users wear "sunglasses with UV protection" and "oven mitts to avoid any burns."

When sold complete with accessories, each SolSource unit retails for about $500. If this sounds like a lot, remember that you never have to purchase fuel when you cook with a SolSource. Ever. Also, the mirrored panels maintain their high reflectivity for six years of heavy use. Another cost benefit: one unit can service more than one household.

These ideas and other clean fuel solutions are now being rolled out at scale, and in a number of critical countries. The effort is being led by an awesome group called the Global Alliance for Clean Cookstoves, based in Washington DC. Their work is critical and necessary. They are saving lives.

So is this the wave of the future? Perhaps.

As I've mentioned before, technologies evolve. Who knows what devices the future will bring? But a simple, reliable cooker that relies solely on the earth's most consistent and free energy resource—the sun?

To me, that sounds like a step in the right direction.

CHAPTER TWELVE

The Shackle of Sewage

For years, global health experts found themselves baffled by a paradox. On average, children raised in India are more likely to suffer from malnutrition than children raised in poorer countries—Somalia, for instance. But how could that be? Shouldn't the child with access to better food enjoy better health? Common sense says yes, indeed. Statistics, however, say otherwise.

The problem seems to compound in light of India's remarkable efforts to feed its poor. As far back as the Bengal Famine of 1943, the government doled out food to its urban poor. Over the years, this effort extended its reach. Throughout the 1970s and '80s, it began to distribute food to more remote areas of the country. Today, the program survives as the Targeted Public Distribution System, or TPDS. Thanks to legislation enacted in 2013, TPDS now assists fifty percent of India's urban population and seventy-five percent of its rural population by procuring food resources.

The system has problems, of course. All systems do. Disingenuous food purveyors substitute government-approved goods with their lower-grade counterparts. Counterfeit ration cards flood locales. Black marketeers withhold the highest quality goods from sale or divert them to higher-paying venues. Overall, however, TPDS has saved millions of Indian lives over the span of generations. It will doubtless save many more in the future.

Today, citizens throughout India can visit more than half a million Fair Price Shops and use their government ration cards to purchase sugar, grain, kerosene, and other staples at below-market prices. At present, TPDS represents the largest food distribution network in the world. International experts have hailed it as a strong, though imperfect tool for combating India's rampant food shortages.

But if it's such a strong program, why are so many Indian children malnourished?

In 2012, the World Health Organization released statistics showing India leading a dismal race. India beat all contenders for its number of stunted, underweight, and wasted children under the age of five. Each of these conditions is caused by poor nutrition, and each incurs correspondingly high mortality rates. Death is hardly the end of the matter, however. Malnourished children who survive past the age of five often lead limited lives. A child's earliest years can be pivotal to his or her future physical and mental development. A deficit of proper nutrients during this window often leads to stunted growth. This, in turn, engenders lifelong cognitive issues, as well as a higher likelihood of contracting diseases such as diabetes and heart disease once the child enters adulthood.

In other words, despite all its efforts to feed its poor—$26 billion in aid per year—India's people are still starving. Again: How can that be?

As luck would have it, the Indian civil rights leader Mohandas Gandhi touched on the answer as far back as 1925. In an editorial for his weekly newspaper, *Navajivan*, Gandhi wrote:

> I shall have to defend myself on one point, namely, sanitary conveniences. I learnt thirty-five years ago that a lavatory must be as clean as a drawing-room. I learnt

> this in the West. I believe that many rules about cleanliness in lavatories are observed more scrupulously in the West than in the East. There are some defects in their rules in this matter, which can be easily remedied. The cause of many of our diseases is the condition of our lavatories and our bad habit of disposing of excreta anywhere and everywhere. I, therefore, believe in the absolute necessity of a clean place for answering the call of nature and clean articles for use at the time, have accustomed myself to them and wish that all others should do the same."

Almost ninety years later, another newspaper verified Gandhi's claim in startling fashion. On July 13, 2014, the *New York Times* debuted an article by reporter Gardiner Harris titled "Poor Sanitation in India May Afflict Well-Fed Children with Malnutrition." The problem with India's starving children, said Harris's article, wasn't that they weren't getting enough food. The problem was that they lived in a country where half the population—some 626 million people—defecated outdoors. (For the purpose of comparison, WHO estimates that fourteen million Chinese defecate outdoors.)

The feces permeates the environment with pestilential bacteria which, in turns, sets off a vicious cycle at the end of which children pay the price.

Many cultures in India accept outdoor defecation as a way of life. Or perhaps it's more accurate to say that they have come to accept it. A week after Harris's article ran, Professor Vamsee Juluri offered a rebuttal in the *Huffington Post*. A professor of media studies at San Francisco University, Vamsee accused Harris of lacing his article with unintentional racism, writing: "The point of the article seems to be this: Hindus don't use toilets because it's against their religion, and it's spreading disease."

Vamsee didn't refute that India suffers from an ongoing

public health crisis. Rather, his article (titled "Nose Deep in Their Own . . . Prejudice: Hinduism and the *New York Times*' Sewage Problem") sought to reorient potential causes of the crisis on the legacy of Western influence in South Asia:

> . . . It is true that the present we live in today is dismal. Hygiene in India is a personal privilege rather than a civic reality. What are the reasons for this? We cannot pretend the last 400 years of colonialism and plunder did not take place and simply say, "This is the way it always is in the benighted third world." . . . Hinduphobia is a reality, not a problem.

I mention Vamsee's rebuttal because, to me, it represents the kind of perfectly misdirected thinking that perpetuates (and possibly encourages) pollution problems worldwide rather than working to eliminate them. It is not inappropriate to be sympathetic to his point of view, but I fear that he's missing the point.

Frankly, it doesn't matter why India has become so polluted with fecal bacteria. Call it the remnants of imperialism run amok. Call it the shackle of sewage left over by white colonial powers. Vamsee's sort of intellectualism will likely inspire plenty of books, papers, and colloquia. But if it doesn't start cleaning the problem up fast, if it doesn't save a single human life, we need to think differently.

A study released by the World Bank in 2010 reported that premature deaths and diseases stemming from India's outdoor defecation problem cost the country more than $50 billion annually. You can say it's an orientalist viewpoint, but I'd like to see this number change.

During my first visit to Bombay, back when I was a student in university, I spent an afternoon playing cricket with some local kids at a traffic roundabout. The game was all about singles and doubles. A boundary meant that anyone hoping to reclaim the ball had to brave the cars whizzing around, a risky enterprise at best. The kids were dirty and shoeless, and much better than me at the game. They clearly loved the fact that a big white guy from Australia was playing with them. Their merciless spin bowling put me to shame.

Halfway through our game, a few of them disappeared down an alley for a few minutes. Later, once we'd finished our round, I continued on my walk. Curious about what those kids had been up to, I ventured down the alley to have a look for myself.

It was the outdoor toilet street. Lined with turds and urine, it was where most of the locals went to defecate. I retreated to another street as quickly as possible. Soon after that, I learned to tell which streets were toilet streets. It was easy, really. I learned to avoid back alleys where men tied up their robes as they came out, and a dust haze had settled over everything, larger than normal. Streets like this were common in all poor neighborhoods. They remain to this day.

Gardiner Harris recounted the scene in his article:

> Millions of pilgrims bathe in the Ganges . . . but a stream of human waste—nearly seventy-five million liters per day—flows directly into the river just above the bathing ghats, steps leading down to the river. Many people wash or brush their teeth beside smaller sewage outlets. Much of the city's drinking water comes from the river, and half of Indian households drink from contaminated supplies.

This sort of thing isn't news, of course. It's been going on for years. The only revelation (perhaps "shock" is a better word) is the new focus adopted so suddenly by so many international health experts, almost in lockstep. In a May 2014 Reuters article by Tom Miles titled "One Billion People Still Practice Open Defecation, Endangering Public Health: UN," Bruce Gordon of WHO summed it up well: "'Excreta,' 'faeces,' 'poo,' I could even say 'shit' maybe, this is the root cause of so many diseases."

Miles's piece goes on to explain how the practice of open defecation spreads cholera, diarrhea, dysentery, hepatitis A, and typhoid, among other diseases. The vector now seems painfully clear. When people defecate openly, they pollute their environment with fecal bacteria, which can infiltrate the intestines of very young children and compromise their immune systems. Infected kids can eat until the cows come home, but they'll never develop properly because the nutrients in their food get shunted off to combat massive and ongoing bacterial onslaughts.

When subjected to such conditions over time, the intestinal lining starts to flatten. According to a 2013 paper submitted by researchers at Boston University, this denigration of tissue decreases the body's ability to absorb nutrients by about a third. In healthy adults, a condition like this could initiate chronic illness. In infants, it can be fatal, or at the very least incredibly damaging.

"When this [diversion of nutritional resources] happens during the first two years of life, children become stunted," noted Jean Humphrey, a specialist in human nutrition at Johns Hopkins' Bloomberg School of Public Health. "What's particularly disturbing is that the lost height and intelligence are permanent."

Based on such a pronouncement, it is appropriate to (respectfully) diverge from the views of Professor Vamsee. Because, as I see it, this kind information—hard science, not to be too specific—is more actionable than a call to sympathize with the current state of Indian culture, from a Hindu perspective or otherwise. Racism isn't the problem; bacteria is.

As it turns out, the problem of outdoor defecation has been linked to other atrocities in India, notably widespread instances of rape. Sex crimes in India stepped into the national spotlight on December 16, 2012. On that date in New Delhi, a twenty-three-year-old female student and her male companion were attacked by six men who stripped them naked and raped the woman repeatedly using an iron bar to inflict massive internal trauma before dumping both victims in the street. The woman eventually succumbed to her injuries. News of the attack broke worldwide, sparking outrage in several countries.

Protestors maintained that violence against women had become a hidden plague affecting many countries, with India chief among them. Dissenters pointed to statistics showing that incidents of rape in India are actually quite low, well beneath those reported by other global regions, such as Southern Africa, the United States, and Western Europe. The statistics may or may not be accurate, but it is clear that the method of rape in India is correlated to outdoor defecation, as indicated by the candid reports that began to surface in the wake of the New Delhi incident.

"When we step out of the house, we are scared," said Guddo Devi, a resident of Katar village in India's largest state, Uttar Pradesh, to NPR in June 2013. At dusk on May 27, Devi's two cousins, young women ages twelve and fourteen, had been kidnapped, raped, and hung by their neck scarves from a tree not far from their home. Both had ventured outside to relieve themselves in a nearby wheat field.

"We [women] have to go [to the bathroom] in the morning, in the evenings, and when we cannot stop ourselves," Devi said. "At times we go in the afternoons as well . . . And there are no bathrooms. We don't have any kind of facility [so] we have to go out."

Devi noted that women in her village frequently traveled to defecate in pairs, hoping to discourage attacks. Her comment appeared to support surveys submitted from around the country by various NGOs. In 2012, for instance, the Swedish organization WaterAid interviewed women in Bhopal, the state capital of Madhya Pradesh. Disturbingly, nine out of every ten women and girls they spoke to said they had experienced harassment while seeking to relieve themselves. Many reported that they had been assaulted.

"A staggering ninety-seven percent of the women we spoke to had not gone to the police to file a complaint, due to feelings of shame and fear of bringing dishonor to their family," said the WaterAid report. "The women were angry at the men who victimized them and frustrated by the authorities and male family members who failed to protect them. Just six percent of women who had been assaulted while going to the toilet received assurance from family members that they would build a toilet in their household."

We might think that all is well and good for the six percent, but this isn't necessarily so. Indications already exist that having more toilets in India might not be sufficient to turn the tide against stunted children or sexual crimes against women.

For one thing, building toilets doesn't mean that people will actually use them. A survey by Diane Coffey, an economist and PhD student at Princeton University's Woodrow Wilson School of Public and International Affairs, of five rural states in northern India found that nearly twenty percent of women with access to toilets didn't use them. Why? Ms. Coffey blamed "common norms, attitudes and beliefs around latrine use" as the culprit. A toilet cannot offer privacy, and therefore safety, when cultural attitudes discourage its use.

Having more toilets might also prove ineffective against the spread of disease. Harris's article told the story of a woman, Phool Mati, who lives in a district equipped with twelve public toilets. Nevertheless, her one-year-old grandson has fallen prey to what appears to be malnourishment caused by the fecal bacteria spread from open defecation. Phool Mati gives her grandson ample food, but he rarely eats and battles chronic illnesses. This confuses her.

"We all use the [toilets,]" she said.

In other words, what's going wrong?

As Harris recounts, the pipe designated to cart feces from the public toilets frequently clogs. At its worst point, sewage overflowed homes in the district, as well as a nearby temple. Use of the public toilet also incurs a fee—two rupees—which might not accommodate a neighborhood's poorer citizens. Harris noted especially that "most children relieve themselves directly into open drains that run along a central walkway."

What can be done? The best answer at the moment seems to have a two-pronged approach: education and invention.

When it comes to education, NGOs have, quite predictably, led the way. Social activist Gouri Choudhury told NPR that her nation's politicians have not yet responded appropriately to the sanitation issue. Ms. Choudhury is a founding member of Action India, a human rights NGO concerned with her country's burgeoning poverty and what her bio calls the "weakening of democratic institutions."

"[Building toilets is] a question of belief in human dignity," Ms. Choudhury told NPR. "Which, somewhere along the line, we seem to have lost."

Ramanan Laxminarayan of the Public Health Foundation of India stuck to numbers when speaking to the *New York Times*. He noted that his government spends more on distributing food than it does on handling hygiene issues by a ratio of sixty to one. Only $400 million gets allocated annually to sanitation improvements. In such a large nation as India, this is a pittance.

"We need to reverse that ratio entirely," Mr. Laxminarayan said.

There are positive indications afoot that the government has begun to heed Mr. Laxminarayan's advice. In May 2014, Narendra Modi, the leader of India's Bharatiya Janata political party, assumed the office of prime minister after a landslide victory against Manmohan Singh of the Congress Party, who had held that office for a decade. While campaigning, Modi had promised to build more toilets. Two months later, his fledgling administration announced that it had budgeted $7.2 billion for the Clean India program, known in-country as Swachh Bharat Abhiyan. The program aims to build toilets for 800 million people by 2019. The spending announced in July 2014 more than quadrupled the previous year's allocation under Singh.

As he presented this budget, finance minister Arun Jaitley struck a sobering note.

"Although the central government is providing resources within its means, the task of total sanitation cannot be achieved without the support of all."

So, yes. Toilet construction ticked upward after Guddo Devi's two cousins were killed in Uttar Pradesh. But India is a nation of 1.2 billion people. I see it as no great stretch to say that no initiative will succeed there unless the government puts its full weight behind hygiene education. Even then, it will be an uphill climb, which is where invention can help.

If there's one thing I've learned as an entrepreneur, it's that old problems never yield to old ways of thinking. You have to start thinking outside the box. You have to try something new or be content with the same old dreary results.

UNICEF understands this, I think. In April 2014, it released an animated music video showing a vaguely Asiatic character whose placid home life gets disrupted quite suddenly by an army of anthropomorphic poop globules.

First the lumps of poop invade the man's dreams. Then he opens his front door to find a life-sized mound of feces stepping on his morning paper. At this point, the Asian man flees. Wouldn't you? But the poop globules chase him on fleet brown feet as thundering techno music rises, the background score to a full-scale fecal uprising.

Look up there! What's that? A giant turd hangs off a skyscraper's spire like King Kong climbing the Empire State Building. Smash cut to a pair of lovers holding hands on a bench in a public park. Their romantic moment is shattered as poop ninjas burst from the shrubs and the flower pots, bouncing up like fleas on crack with white grins lighting their dark brown faces.

The Asian man steps in a lump of feces while walking down a sidewalk. When his shoe divides the lump in two, a jubilant cartoon banner explodes. HAPPY BIRTHDAY! it says, and the meaning is clear: shit is resilient. It multiplies. Now there are two lumps where one used to be. When the man tries to shake both lumps off his shoe, they cling to his foot with ferocious tenacity as poop lumps start forming conga lines that jig through the streets of a horrified city.

Have you started to feel disgusted yet? If you have, you're not alone, and I'm sure this reaction is exactly what UNICEF wants you to feel. Perhaps now you can see why I admire their way of thinking. Some people try different tactics, sure. A precious few will break the mold. But only the rarest and most dynamic social entrepreneurs will commit to something so likely to fail.

Now, let's examine the results.

The UNICEF poop cartoon was so wacky, so off-color and, well . . . disgusting that it was sure to go viral. Which it did. And viral videos get people talking.

Did you see the cartoon about the . . . well, the *shit*?

They wanted to start a conversation, of course. And they succeeded. The revulsion people felt toward the cartoon is the same revulsion that built a new and fairly indelible public meme. This meme makes it clear that certain countries have grown deeply polluted by their own people's feces. It hints that these countries aren't safe to visit. That you can't find peace or happiness when you live or visit there.

At this point, government ears start to perk up. Politicians begin to interject carefully crafted messages into their speeches and public statements. Why? Because in this era, no one can afford to have their country ostracized from the world community. The moment a meme starts to undermine a nation's GDP, you can bet someone in power will seek to correct it, or he or she won't be in power much longer.

On a larger scale, the UNICEF poop cartoon employed the same tactic I once used against the manager of the Tanzanian toothpaste factory, and so many other polluters after that. I'm not a big proponent of using shame to correct behaviors with my friends, family, or business associates. However, in larger arenas, like the behavior of nations, it can be effective.

The question is, will it be enough to help the sick children of India?

Some social entrepreneurs aren't willing to wait on this score. I'm thinking now of Bindeshwar Pathak, the noted Indian polymath, social entrepreneur, and author of *Road to Freedom: A Sociological Study on the Abolition of Scavenging in India*.

Pathak rose to prominence after choosing to live among India's scavenger clans while pursuing his doctorate in sociology. The experiences he lived through and wrote about resonate deeply with those more recently reported in Katherine Boo's book *Behind the Beautiful Forevers: Life, Death, and Hope in a Mumbai Undercity*. Convinced that true change could not be achieved by traditional social apparatuses, Pathak established Sulabh International Service Organization, a nonprofit dedicated to the "Gandhian ideology of emancipation of scavengers" and other social causes (according to the Sulabh website).

Recently, Sulabh took on sanitation issues among India's untouchable class. The organization designed a two-pit, pour-flush, composting toilet it calls the Sulabh Shauchalaya. It totes the Sulabh Shauchalaya as a "socially acceptable, economically affordable, technologically appropriate [toilet that] does not require scavengers to clean the pits and [has been] implemented in more than 1.2 million houses all over India." Doubtless for this reason, many media outlets in India have dubbed Bindeshwar Pathak "The Toilet Guru."

The design for the Sulabh Shauchalaya is simple. Picture a toilet that users flush manually by pouring a modest amount of water, say a liter or two, through a trough. The water sluices waste down a pipe that terminates at a Y-intersection. Each pipe stemming off the Y leads to a pit lined with honeycomb-surfaced bricks. Only one pit collects waste at a time. When the first pit fills (which should take about eighteen months or so, depending on the size of the household), the homeowner blocks it off and lets it sit while opening the pipe to the second pit. Filling the second pit should take the same amount of time: about another eighteen months.

By the time the second pit fills, the feces in the first pit will have composted itself and can be safely repurposed as manure fertilizer. Having two pits also prevents household members from coming into contact with fresh and therefore bacterially dangerous feces. If they follow a fairly obvious schedule, they can empty a pit full of dried manure, switch tanks, and the process repeats itself. The newly filled pit sits for the next year and a half. Its feces dries out, composting itself. Repeat. Repeat. Repeat.

Each Sulabh Shauchalaya costs between $100 and $1,000 to build. Sulabh outfitted each of the 144 households in the village of Hir Mathala in the state of Haryana with a unit for about $250. The program has been successful. Villagers have reported better health since the toilets were installed. Women, in particular, have offered their thanks.

"There has been a huge change in our lives," said Vijay Laxmi, a thirty-five-year-old mother. "Before [we had the toilets], the men would follow us, wait for us to sit in the field and watch. Now, thanks to Mr. Pathak, we have a lavatory at home. [Now] we don't need to step out, and we feel better. Our dignity, which is an ornament for us—is now safe."

All that is well and good, but again, let's look at the numbers.

While I hate to burst anyone's bubble, the sad truth is that Bindeshwar Pathak's nonprofit has spent the past forty years building 1.3 million toilets across India. In all that time, after all those flushes, his country's rate of open defecation hasn't altered appreciatively. If anything, it continues to rise while the health of its people continues to fall.

We need to get our arms around this problem, and fast.

CHAPTER THIRTEEN

Atomic Toxins

Late April. A quiet Friday night in a typical mid-sized city. Citizens in bed, after a normal day like most others. Some had gone to work, some had gone shopping, and kids had gone to school. It had been a normal day, a good day. The night was clear and cool.

Then, in the distance: the sounds of explosions, two in rapid succession. It was one o'clock in the morning and the fire alarms started to ring, calling the men to the fire trucks. In the distance, the red blaze from a burning building lit up the night.

By morning, rumors were spreading fast—something about an accident at the power station down the road. There was an announcement that the national government had already been notified. Already, a commission had been established to study the problem, which made the locals suspect the worst. Establishing a commission was one of the national government's time-honored tactics.

National troops arrived in the town square late that evening. They were led by a prominent doctor from the Academy of Sciences. Within hours of their arrival, at three o'clock on Sunday morning, a voice began booming through loudspeakers that the soldiers had set up in the streets.

"Attention, residents! Your city council informs you that,

due to the accident at Power Station, radioactive conditions in this vicinity are deteriorating. The Ruling Party, its officers, and the armed forces are taking necessary steps to combat this. Nevertheless, with the view to keep people as safe and healthy as possible—the children being our top priority—we must temporarily evacuate the citizenry . . . "

Right on cue, buses began to arrive at the city's various apartment buildings. The voice on the loudspeakers kept talking. It told everyone to pack important documents, valuables, and a bit of food before leaving their homes. It said:

"Do it quickly! There is no reason to worry. Repeat: no reason to worry! Residences will be guarded by local police and the National Guard. Meanwhile, if everyone could please remember to turn off your lights and electrical equipment before you leave their homes, that would be very much appreciated. And would you also close the windows, please? Thank you so much. Now hurry!"

People complied. What choice did they have? They were confused. The soldiers had guns. Officers of the National Guard answered anyone who questioned them with talking points:

"Remain calm."

"All is well."

"This evacuation will be short-term."

"Very likely it will only last three days."

By the end of that Sunday, more than 100,000 people had evacuated the typical, mid-sized city and its suburbs. As time went on, another quarter of a million residents had to be moved from the outlying regions.

Today, almost thirty years later, only a handful of people have returned to the general area. No one whatsoever has returned to the typical, mid-sized city, and with damn good reason.

That city was Pripyat, located on the northern border of what is now Ukraine's Kiev Province. The date was April 26, 1986. And the power station down the road—three kilometers, to be precise—was the Chernobyl plant, site of the worst nuclear disaster in history.

The Chernobyl meltdown emitted more than 400 times the radioactive material released at the bombing of Hiroshima. It was a Level 7 event on the International Nuclear Event Scale. Level 7 events comprise the direst end of the INES scale. They are categorized by a major release of harmful radionuclides that, in turn, create a catastrophic impact on human health and the environment.

Chemically speaking, radionuclides are atoms that contain unstable nuclei. The instability causes the atoms to decay, the process of which, in turn, emits high levels of radioactivity either as gamma rays or subatomic particles. Even at low levels (which was not the case at Chernobyl at all) these rays or particles damage the cells of living tissue, affecting the DNA in ways that exacerbate cancer.

The real tragedy of Chernobyl is how easily its disaster could have been prevented. Following the accident, Soviet officials lobbied intensely to blame the incident on operator error and/or design deficiencies in the reactor itself. But independent analysts found that neither version was true. According to a report released in 1992 by the International Nuclear Safety Advisory Group:

> The accident [at the Chernobyl nuclear power plant] can be said to have flowed from a deficient safety culture, not only at the Chernobyl plant, but throughout the Soviet design, operating and regulatory organizations for nuclear power that existed at that time.

On that fateful night in April, Reactor 4 at the Chernobyl plant was being taken offline for what should have been a routine systems check. But proper calculations weren't made to compensate for how hot the reactor cores remain during shutdown. When the reactor output spiked, creating an unexpected power surge, the plant's emergency shutdown equipment failed to compensate. This triggered a power surge that ruptured Reactor 4's containment vessel, setting off a fusillade of steam explosions that blew the roof right off the facility and sent a column of radioactive smoke rocketing into the atmosphere.

Fires raged for ten days while the atomic cloud expanded over much of Russia and Eastern Europe. Wide swaths of Belarus and Ukraine became heavily contaminated. The cloud's diffusion makes it difficult to measure its precise impact on human health and the environment. Without question, however, emergency personnel who responded to the explosion at Reactor 4 bore the brunt of the trauma. In fact, because they make up such a closed group, these men and women present some of the most measurable data we have about the incident's long-term impacts to human health.

According to reports, 600 people were present at the reactor site either at the moment of explosion or afterward, during the containment and cleanup processes. Within this group, forty-one deaths occurred in total: two from the plant's explosion, four in a helicopter accident, and thirty-two from acute radiation syndrome. Almost all ARS fatalities expired months after the incident. Over long weeks, the radiation they'd been exposed to ate them alive.

Within the overall group of 600, another 137 cases of ARS were reported, but people in this subset managed to survive despite having received consistent doses of up to 800 roentgens. (A dosage of 500 roentgens in five hours is usually more than sufficient to kill the average human being.)

As I mentioned, however, beyond this data set, few concrete metrics exist. Politics are partly responsible for this. No one knows for certain how accurately the former Soviet regime reported on the catastrophe.

Another report was released by the World Health Organization in 2005. This painstaking document outlines the Chernobyl disaster's likely legacy, including many far-reaching effects that may yet come to pass.

For instance, stemming from the incident:

- Some five million people throughout Belarus, Russia, and Ukraine have likely been contaminated by radionuclides.
- We could see as many as 2,200 more deaths caused by exposure to radiation released at Chernobyl.
- Relocating more than 100,000 people from the incident's highest impact zone did little to reduce their exposure to harmful radiation.
- More than 350,000 relocated citizens suffer deeply from the psychological impact of their displacement. This syndrome, which the report calls "paralyzing fatalism," has been described as markedly more harmful than physical exposure to radiation.

Finally, the report presented these two shocking data points:

- The sarcophagus which engineers hastily built to contain the ruptured Reactor 4 has "degraded," creating real danger of further exposure to radioactivity.
- No concrete plan exists to clean up the Chernobyl site.

This last point, which bluntly highlights the risk of further exposures, has galvanized the Ukrainian government and the international community to build another containment vessel on top of the first one. At a cost of almost $1.5 billion, a new building will soon roll over the top of the old reactor, hoping to keep its contents safe for the next hundred years.

Today, almost thirty years later, Pripyat has become a ghost town. Letters, stamped but never sent, paper the floor of the once bustling post office. In a local amusement park, bumper cars that ground to a halt on the day of the accident stand still as gravestones, slathered in decades' worth of dust and debris. Trains that once ferried people and freight from town to town squat on rails overgrown by tall grass and shrubs; their metal frames, turned brown with rust, have started to run like wax sculptures under a heat lamp.

The economic costs of Chernobyl are staggering, and they continue to grow. The USSR virtually bankrupted itself by spending the equivalent of $18 billion (in 1986 terms) on mitigating the disaster. Belarus has spent another fortune on projects related to Chernobyl's aftermath. Neither of these figures accommodates the costs of economic displacement such as the short- and long-term results of collapsed worker productivity, impacts to industry output, the toll on the health care industry, the loss of agricultural output from heavily contaminated zones, and so on.

If all this sounds bad, Chernobyl isn't the only radionuclide disaster in this part of the world. Far from it.

In the years following World War II, the Soviets raced to develop their own atomic bomb to match that of their archrival, the United States. They built several secret factories to generate weapons-grade plutonium and uranium, and pushed their workers very hard, neglecting basic safety precautions. The old phrase "haste makes waste" acquired a new and deadly connotation.

In the Ural Mountains of southwestern Siberia, a facility called the Mayak Chemical Combine dumped untold quantities of medium to high-level radioactive waste between 1949 and 1956. Mayak sported six nuclear reactors whose construction was based as much on stolen intelligence and supposition as it was on science. Inefficient and dangerous, these reactors leaked much of their waste into a local river, the Techa, whose water, fish, and surrounding wildlife fed many nearby communities.

The Techa is but one leg in a 1,000-kilometer-long river system that empties into the Arctic Sea. Contamination from Mayak has made its way to the ocean and been detected there.

The Mayak facility's leaking nuclear waste disposal facilities are bad enough. But on September 29, 1957, a cooling unit at the enrichment plant failed, causing an explosion that blasted about eighty tons of radioactive detritus into the atmosphere. It was a Level 6 event on the International Nuclear Event Scale—the third worst event of its kind in history, behind Chernobyl and Fukushima Daiichi. The toxic cloud released at Mayak contained twice the amount of curies Chernobyl would emit some thirty years later. Eventually, the cloud expanded to cover some 24,000 square kilometers over three Soviet provinces, while irradiating tens of thousands of innocent people.

Don't kick yourself if you've never heard of this event. Back in the day, Soviet officials considered the Mayak plant so top secret, it didn't show up on a single contemporary map. Everything I've just told you came to light many years after the event took place. During all that time, however, a number of people have had to live with the specter of Mayak's toxic legacy.

Thirty kilometers downstream from the plant, death has become a way of life in the village of Muslyumova. One child in four is born with genetic mutations, and less than two percent of the population can be called clinically healthy. Workers who live in this region rarely live to retirement age. At one point not too long ago, seventy percent of the locals had been diagnosed with leukemia. Cancer rates have spiked to as high as five times the Russian national average, with an especially marked increase in cancers of the digestive system, bones, and lungs.

Some studies show that half the men and women of childbearing age are sterile. Rates for child morbidity and mortality have reached three times the national average. The numbers for miscarriages and babies born prematurely are off the charts. And thirty percent of all babies born have physical disorders or defects.

As perhaps the ultimate insult, Soviet officials denied that anything had happened at Mayak. Instead of offering aid to the victims, they somehow legislated that all the surrounding villages in that region could be evacuated *except* Muslyumova.

The village's 4,000 residents appealed this decision repeatedly, begging to be resettled. Repeatedly, their pleas were ignored. And yet, Soviet officials cared enough to make the citizens of Muslyumova undergo mandatory blood and bone marrow testing from 1950 forward.

For three generations, awful rumors abounded over what those tests were intended to show and why no one in Muslyumova had ever seen the results. More than a few citizens decried being used as lab rats in some bizarre science experiment. The Muslyumova data was finally declassified in 1992, shortly after the fall of the Soviet Union. They showed that radiation had indeed afflicted the population for years, fettering them with ailments and diseases that eventually led to untimely death. These results seem to correspond with tests conducted by independent assessors who measured quantities of Cesium 137 between 300 and 500 nanocuries per kilogram in the Techa's silt bed—high enough concentrations to qualify as solid radioactive waste.

A few legislatures local to that region have passed resolutions that the villagers of Muslyumova should be compensated for their suffering, that their town should be relocated to safer confines. But money is always at issue in that part of the world. To date, nothing has been done to make good on these notions.

A third radionuclide disaster took place in the Fergana Valley, an agricultural cornucopia wedged between the towering Tian Shan and Gissar-Alai mountain ranges in Kyrgyzstan. The ethnic diversity of Fergana's inhabitants punctuates the valley's history and geography. Eastern and Western civilizations first collided here more than two millennia ago when Greek, Persian, and Chinese traders exchanged goods in the markets of local villages. The Russian Empire subjugated the land during the 1800s; decades later, the Fergana Valley became part of the Soviet Union, at which point its troubles began.

Near the top of the valley, on the bucolic shores of the Maili Suu River, the Soviets built another of their infamous super-secret and unsafe uranium processing plants. Between 1946 and 1948, the Maili Suu facility processed about 10,000 tons of uranium ore, which was used to construct the Soviet Union's first atomic weapons. The leavings of this industry created twenty-three pits and thirteen dumps used to capture tailings from uranium mines and processing operations.

Positioned on a hillside above the river, these pits and dumps contain an estimated 2.5 million cubic yards of radioactive waste, which plant officials buried under thin layers of gravel and topsoil. This would present a dangerous situation in and of itself. But this particular region of the Fergana Valley is also considered tectonically unstable.

To quote an October 2000 article from the *New York Times*:

> . . . Landslides and floods . . . could send the [radioactive] material into the river and ultimately the water system of the valley. Just as winds spread contaminated dust from the Chernobyl nuclear plant accident in 1986, the rivers, streams, and irrigation canals that lace the valley downstream could carry radioactive materials throughout the 60,000 square miles in the basin between the Tian Shan and Pamir-Alai ranges.

By the *Times*' assessment (based on research submitted by the United Nations Development Program), the situation at Maili Suu had the potential to "threaten the lives and livelihoods of ten million people in three Central Asian countries" (Kyrgyzstan, Tajikistan, and Uzbekistan). One senior Russian official disclosed that the site's contamination could "destroy the agriculture base [of the Fergana Valley], force the immediate evacuation of 500,000 people and damage the economies and stability of all three countries."

Sadly, the *Times* prediction eventually came to pass. In the spring of 2002, a massive mudslide altered the course of the river, creating new lagoons that came perilously close to barrows filled with toxic waste. Then, three years later, an earthquake struck. The resulting landslide sent an estimated 300,000 cubic meters of highly radioactive material into the river. Presumably these materials were carried downstream toward civilian populations.

Conditions warranting a full-scale emergency were basically a matter of time. But the classic question remained: Who would lead the charge to clean up somebody else's toxic mess?

Pure Earth has been doing its part as much as its resources will allow, conducting cleanup efforts at all three sites I've listed above. We will likely continue to take more project work of this sort in the near future. Our European partner, Green Cross Switzerland, has been instrumental in helping us with these projects.

Cesium 137 was the most abundant contaminant released by the Chernobyl meltdown. External exposure to this element can cause severe skin burns. Once ingested, however, this radionuclide spreads through the victim's soft tissues and muscles, causing cancer via gamma radiation. Cesium 137 has a half-life of about thirty years. This means that about half the quantity of this radionuclide released from Chernobyl will continue to affect every landscape it has touched until 2016, give or take. That's too much pollution at work for too long, and that's why we started our project work in Bryansk Oblast.

At first blush, people might wonder what Bryansk has to do with Chernobyl. The province is located about 110 miles from Reactor 4. Could it really have sustained much damaged? The answer is yes. Remember, the toxic cloud that emanated from the explosion covered most of Europe and Russia. In both cases, many regions received fallout levels that, while not quite serious enough to warrant permanent evacuation, nonetheless proved sufficient to irradiate local ecosystems and human populations. Bryansk was chief among them.

A cradle of Russian agriculture and manufacturing, Bryansk

received heavy doses of Cesium 137 over approximately two million acres. An estimated 500,000 people live and work within the affected confines. Even now, they run a very high risk of ingesting radionuclides that have accumulated in meat, milk from local dairy animals, and produce from local farms.

When we first arrived in Bryansk, we were well aware that over twenty percent of dairy milk had become dangerously contaminated in many Russian provinces. Children, of course, formed the highest risk group, not only because of their size and the relative vulnerabilities their developing bodies presented, but because children tend to consume more dairy products than adults do.

In collaboration with our local partner, the Veterinary Laboratory of Bryansk, and alongside Green Cross Switzerland, we purchased chemical sorbents—materials that, by nature, absorb specific liquids and gases. If you cracked open the breathing mechanism on a gas mask, for instance, you'd find rows of sorbent materials arranged in such a way as to provide maximum filtration against toxic gases. The idea was to put a particular sorbent to work in rehabilitating local food supplies. That sorbent was Bifezh, a hexacyanoferrate compound. Beginning in September 2005, we added 1.5 tons of Bifezh to feed consumed by Bryansk dairy cattle.

This amount of sorbent should have been sufficient to clean about 350 tons of milk the animals produced. Did it work? Indeed! Quite well, in fact. We were pleased to watch radioactivity levels in treated milk drop by over ninety percent, rendering it safe for human consumption.

Actions of this sort did nothing to reverse the deleterious effects of radiation previously absorbed by the population. But they significantly reduced the burden of radionuclide absorption moving forward.

Our pilot program was such a success, authorities began to examine how they might repeat it on a much larger scale. Pure

Earth specialists estimate that, by employing twenty tons of sorbent on an annual basis, we can purify over 5,000 tons of milk and 1,500 tons of meat each year.

That's the sort of intervention I like: easy to wrap your brain around, simple to implement, relatively inexpensive, and cumulatively beneficial. If only all of our projects could follow a similar template.

Helping at the Mayak site took a much different path. First, there were problems between local authorities and NGOs, with each claiming different results from their tests. Because of this, remediation efforts had gone nowhere. Our first plan was to harmonize the efforts of all parties. We wondered how much good they could accomplish if they stopped harassing each other and synchronized their efforts instead.

We donated research-quality equipment so that all groups could begin taking routine, reliable measurements of local radioactivity levels. NGOs and government groups were able to calibrate their instruments. They then began meeting at regular intervals to compare their findings. How much radioactivity had accumulated in the local soil? In foodstuffs? In dairy products? And so on.

The results of this pilot project were eventually published in an international scientific journal. This, of course, called attention to Muslyumova's plight, but we wanted to do something tangible, so we plowed ahead with a program to remove radioactive sludge from the riverbanks of the Techa.

We were particularly shocked to find that soil and sediment sampled from the riverbanks serving as the town's swimming hole showed massive deposits of harmful radionuclides. So Pure Earth did what Pure Earth does best: we rolled up our sleeves and got to work. With the help of a local partner, we bulldozed the irradiated mud, scooped it up, and hauled it to a suitable landfill site where it could be properly contained. We then rebuilt the riverbanks with good clean soil that could be monitored periodically to ensure no future contamination.

Our field studies estimated that this effort impacted the lives of 124,000 people. Of course, that's just the beginning of what we hope becomes a much larger effort to clean up this region in the future.

In the Fergana Valley, Pure Earth harkened back to a 1999 study conducted by the Institute of Oncology and Radioecology, which showed that cancer rates throughout the area had doubled the country's norm. Frankly, this came as no surprise since uranium is a known carcinogen. However, as it decays, uranium also produces radon gas, which experts call the most prevalent cause of lung cancer apart from smoking.

Baseline radiation samples we took in local schools and hospitals proved dismal: the places were highly toxic. So we initiated a schedule of interventions. First, we upgraded each schools' water filters, and when situations warranted full renovation of water supply systems, we provided technical and logistical assistance. We also created an educational series: training seminars for school and café employees, and educational sessions for school children. The idea was to give everyone a crash course in anti-radiation hygiene techniques, which include aggressive hand washing, dust abatement through sweeping and rinsing, proper use and care of water filters, and the shredding of meat and vegetables before storage (this allows the foodstuffs to be more thoroughly washed and helps remove radioactive contaminants).

During the course of these interventions, we discovered two families whose dwellings had become so severely contaminated by radon that we evacuated them immediately. More families await relocation as of the writing of this book. Based on these findings alone, we began passing out radon counters so we could track this particular thrust of the crisis and intervene as needed.

Levels of uranium and other contaminant metals in the drinking water we tested at schools and hospitals dropped by forty-eight to sixty-five percent post-intervention. We also noted that in-room exposure to residents of all ages decreased by thirty-eight to fifty-five percent. This exposure was mostly caused by radioactive materials embedded in walls and/or radon gas seeping up from the ground. In each case, mitigation was required. All this work was done, once again, with Green Cross Switzerland—good guys, indeed.

With all three radionuclide projects, we obtained good results from solid, boots-on-the-ground-style projects. But so much more remains to be done.

The methodology for combatting radionuclides will likely sound the same to you as our efforts to tackle other pollutants: devise and implement a pilot program, prove its effectiveness at reducing the pollutant at hand, then seek out sponsors who can fund an expansion of efforts to cover a wider and deeper scale.

This is what we're doing now throughout the former Soviet Union. Considering the scope of the problem, we'll likely be at it for some time.

CHAPTER FOURTEEN

The Scourge of Hexavalent Chromium

In my business, it's relatively easy to talk about lead and mercury poisoning, air pollution, and pesticides, since most Westerners possess a baseline understanding of these issues. For years, our governments have warned us to avoid leaded fuels, lead in paints, lead in basically any incarnation. Our parents told us never to break thermometers since the mercury they contain can be bad for us. We've read about smog affecting London, Los Angeles, and Beijing. Our documentaries have expounded the crisis of greenhouses gases. And so on.

Sadly, few people have heard about hexavalent chromium and the dangers it poses. Considering how ubiquitous the stuff is, we need to rectify this, and immediately.

A naturally occurring metallic element found in water, soil, and rocks, chromium is something we encounter all the time. Virtually all foods contain chromium; we normally ingest it through meats, mollusks, crustaceans, produce, and unrefined sugar. And it's good for us. Chromium regulates our ability to metabolize sugars and maintain healthy reactions between enzymes. In fact, chromium is so important for our bodies that the U.S. FDA has set guidelines: adults should take between fifty and 200 micrograms of chromium a day to maintain optimum health.

So chromium's great, right?

Not so fast.

The chromium referenced above is a common form of the element known as trivalent chromium, or chromium 3. But what happens if you oxidize the element further, say, to adapt it for industrial purposes? Take away three more electrons, and trivalent chromium becomes its much more dangerous cousin, hexavalent chromium, also known as chromium 6, chromium 6 +, or often quite simply as "hex."

Hexavalent chromium is used all over the world to catalyze many industrial processes. It shows up in leather tanning, textile dying, wood preservation, surface coating to prevent corrosion, and the production of stainless steel. It's a key ingredient in many paints, plastics, inks, and dyes. It's often a byproduct of the welding process. And it can be lethal. Once ingested, inhaled, or absorbed through the skin, hexavalent chromium acts as a mutagen, which means it literally recodes our DNA to invite diseases, notably cancer.

Inhaling hexavalent chromium can lead to irritation of the respiratory system and gastrointestinal bleeding. Ingesting hex will likely cause stomach upsets, ulcers, and severe damage to the kidneys and liver, as well as premature dementia. In both cases, death can occur after a certain level of exposure.

So the difference between chromium 3 and chromium 6 is huge. Where chromium is concerned, three is company, but six can make us sick.

To introduce the dangers of chromium 6, I'll highlight one industry whose worst practitioners notoriously abuse the metal: the leather tanning industry. But before you leap to any conclusions, please note that I'm singling out the industry's *worst* practitioners. The scenarios I'm about to detail have very little, if anything, to do with A-grade leather manufacturers.

Many Western firms are justifiably famous for their leather products—shoes, furniture, handbags, garments, jackets, wallets, boots, and belts. You probably recognize the retailers involved in this sphere. Nike, Gucci, Coach, Chanel, Prada, Kenneth Cole, Versace . . . the list goes on and on. For the most part, however, these firms go to great, even admirable lengths to secure the best products from the most reputable dealers whose environmental safety routines have been well vetted. On the whole, we do not need to worry about this group. It's the lower standard, the B-level practitioners, that cause problems.

It's counterproductive to blame big Western conglomerates who, by and large, do a much better job at policing themselves than the public gives them credit for. These firms are rarely the culprits when a pollutant gets abused. If anything, we should ask for their help in dealing with their lower-on-the-food-chain brethren, who create the actual issues.

To paint a picture of these bad actors, let me first explain what tanning is, and what the process entails.

Tanning refers to any process used to treat an animal skin so it becomes its more durable and workable counterpart, the item we know as leather. Some tanning techniques are as old as human civilization. Ancient peoples used dozens of methods. American Indians would stretch a fleshed hide around tree trunks before rubbing it down with a paste made from the slain animal's brains. The Romans, I'm told, gathered buckets of urine from communal toilets and soaked their skins in it to soften them and remove excess hair. Some cultures rubbed their hides with feces to get the same effect. Others let them putrefy for a matter of months before soaking them in salt solutions.

In a modern tanning facility, workers receive raw animal hides. These hides are covered with hair or fur on one side. The other side contains residual fat deposits, meat, blood vessels, and so on. Both sides get scraped until the skin is all that remains. The skins are then placed in processing drums, which are large, often measuring four feet in diameter. More skins are added to create a load, just as people often fill a washing machine to do laundry. Tannery workers then add water to the drum, as well as certain chemicals like biocides to prevent bacterial growth, degreasers to remove residual fat, and sodium hydrosulfide to remove any lingering hairs. But the most important chemical additive is trivalent chromium, which interacts with the skin to make it soft and pliable while increasing its resistance to heat, making it less susceptible to rot and stabilizing its color.

Trivalent chromium also hastens the tanning process. The drum begins to rotate, the skins slosh around in the chemical slurry, and within a day or so, they are done. A tanner then retrieves the skins, which have turned light blue from the chromium (which is why, in industry terms, you hear such skins described as "blues" or "wet blues"). Think of this as the beginning stage of the more complicated tanning process. From here, the skins can be treated further to produce whatever effect the tanner would like to achieve.

At present, approximately eighty percent of leather worldwide is produced using chromium 3. Which is fine. Trivalent chromium isn't something we'd want to drink out of our coffee mugs, but the case for its use in leather tanning is sound.

Does this chemical help an industry become dramatically more profitable? Yes.

Is it more effective than any known alternative? Yes.

Can the chemical be handled quite easily and safely? Yes.

Is the chemical recyclable? Yes! Top-grade tanners recycle their chromium all the time by treating their residual slurry in specialized treatment plants. The recycling process negates any impact the chromium might create on the environment while yielding cost savings for the manufacturers—a win-win situation.

So trivalent chromium is generally fine, provided that it's handled responsibly. As I mentioned, however, the chromium scourge isn't perpetuated by responsible tanners. Low-grade tanners exist all over the world and, like low-grade practitioners of any art, they're always looking for shortcuts, which is where the problems begin.

Mom-and-pop tanners in low- to middle-income countries rarely recycle their slurry. They dump it into the local river instead, or in the empty dirt lot next door to their facility. In both cases, the chromium 3 permeates the soil where, through a fairly simple and well-established oxidation process, it converts into deadly hex.

Other low-grade craftsmen substitute hexavalent chromium for trivalent chromium in their tanning process. Why? Because hex can be purchased for about the same amount of money as trivalent, and hex reacts more quickly with raw hides. The tanner's facility can produce more wet blue within a specific period, and this increased volume of product translates to greater profit.

But in either case, we end up with the same situation: large pockets of trivalent and hexavalent chromium infiltrating groundwater tables at very dangerous levels. Which leads to hex showing up in people's drinking water. Which leads to diseases.

Hexavalent chromium can blister skin on contact. When people drink hex, this same reaction occurs in their innards. It should therefore come as no surprise that people who drink hex end up developing cancer and dying horrible, painful deaths.

India and Bangladesh feature some of the largest, most industrious tanning centers in the world, and also some of the worst. In both countries, low-grade tanning facilities frequently comprise the backbone of local economies. Both countries feature numerous legacy contamination sites, where tanners have polluted the environment with chromium for generations on end. Both countries also host manufacturing facilities that produce chromium for the tanning industry. These same manufacturing facilities are often guilty of improper waste disposal.

The Indian city of Kanpur, located a couple hours' drive due east of Delhi, is the second largest and most populous urban economic center in northern India, the ninth largest city in the country, and the largest by far in its home state of Uttar Pradesh. Kanpur is also one of the most polluted places in India. Its eastern districts feature some 350 industrial leather tanneries, many of which discharge their untreated waste into local groundwater sources and the Ganges River. This waste contains thoroughly toxic levels of chromium, mercury, and arsenic.

In 2005, I visited Kanpur with Promila Sharma, Pure Earth's country program coordinator for South Asia. Funding for our venture came from the Asian Development Bank, an organization founded in 1966 with a mission to fight poverty in Asia and throughout the Pacific. The first and most important stop on our itinerary was Kanpur's Noraiakheda district, a settlement of about 30,000 people.

"This is a bad situation," Promila said, as our driver approached the village. Normally ebullient, her manner had shifted abruptly to a sort of distracted calm. By this point, I had seen many highly polluted neighborhoods and had learned to brace myself accordingly. I found Promila's reaction sobering. A native of India, she had witnessed some of her country's worst pollution scenarios, but I'd never seen her react like this.

The Noraiakheda settlement had developed within a large plume of hexavalent chromium emitted by a pile of toxic waste. A chemical plant that once supported the local tanners had dumped its effluents into its own backyard. Piles of sludge were speckled with puddles of yellow water—a telltale sign of chromium contamination. The plant had been closed, but no cleanup had been done on the sludge at that time. Worse yet, as rain leached the chromium into the groundwater, it provided the proper circumstances to oxidize the chromium into hex.

Standing on the outskirts of the field, we watched small

children playing in the mud and young mothers shuttling this way and that with one, sometimes two babies slung on their backs. Their bare feet sloshed through the waste, splashing mud up their legs and onto their clothes, while toddlers stomped through the puddles with glee. Slum houses were moving in on the area, and people were living right on top of it all.

"It gets worse," Promila said. "There are pockets of methane trapped in the sludge. During the hot summer months, the pockets catch fire."

"And people get burned?" I asked.

She shook her head. "Rarely. But the fumes the fires generate are deadly. Each conflagration releases more toxins into the air. Chromium that hasn't soaked into the groundwater gets inhaled."

"Where does the village get its drinking water?" I asked.

"Kanpur has installed several water treatment plants," Promila replied. "They have not been successful, I'm afraid."

"How bad?" I asked.

"India's national government set the limit for chromium levels in drinking water at .05 milligrams per liter," she said. "But the Central Pollution Control Board conducted a study in 1997. The readings here and elsewhere in Kanpur . . . "

"Yes?"

"They hit 6.2."

I took a moment to absorb this information. "Okay," I said finally. "What do we do?"

Promila frowned. "At the moment, I'm not really sure," she said.

Once again, it was time to start educating ourselves.

We knew that we wanted to create a pilot project aimed at lowering the hex levels in people's drinking water. We also felt there was little we could do about the contamination in the Ganges; the problem there was so vast that it was beyond our immediate control. But education would help—getting the message out to the locals so they would know what was dangerous. We hired a local Kanpur NGO called EcoFriends to generate messages and educational programming that would warn locals about the river's toxicity and advise them on where to draw their water. This placed our focus on bore wells, which we knew we would have to clean up.

To do this, we partnered with Kanpur's wing of the Indian Institute of Technology, a national university program specializing in public engineering and management education. We also tapped experts from our stable of U.S. environmental engineers, and specifically those who had experience with water tables permeated by hexavalent chromium. We also worked closely with the Central Pollution Control Board's North Zonal office, a federal regulatory body and pollution watchdog. CPCB coordinated experts and the pilot trial of what became our methodology.

Eventually, we learned that the U.S. EPA follows a standard process for situations like the one in Kanpur: First, you set up some pretty sophisticated equipment and get to work pumping out all the groundwater. Second, you run the water through a cleaning process. Third, you return it to the subterranean tables where it originated. Then you repeat the process. Pump, filter, return. Pump, filter, return.

Promila frowned when she heard this methodology described.

"I understand that it's a tried and true method," she said. "But I foresee a number of factors that will make it prohibitive in India."

She was right. For one thing, the EPA process would have to be reiterated almost constantly over several years. But the toxicity levels at Noraiakheda were so great, we feared that thousands of people would die in that interval. We wanted to effect change faster.

"The EPA technique also requires a dedicated, stable power

supply, which we won't have," Promila pointed out. "Also . . . how can I put this?"

"Just say it," I said.

"I predict that certain enterprising members of the affected communities will make an art form of stealing and selling whatever high-tech pumps we position throughout their area."

I nodded. "There are cost prohibitions, as well," I said. "The EPA method costs millions of dollars that we simply aren't budgeted for. We need an alternative that fits our confines and resources."

Everyone on the project agreed. It looked like our process had stalled for the moment, but a few days later, our team of experts came up with an idea. Promila laid out the details of this new plan, which would begin with a process of in situ bioremediation. "In situ" is a Latin term that literally means "on-site" (as opposed to "ex situ," which means "off-site"). Those in the business of soil remediation generally prefer in situ techniques because they rely on a process of adding something to a contaminated site, whereas ex situ jobs rely on taking something away. Ex situ work, for example, could involve excavating land by the ton, hauling it someplace for treatment, incinerating it if necessary, and so on.

This new plan would convert the hexavalent chromium to trivalent chromium while it was still in the earth.

"In this case," Promila said, "ex situ won't work because of—"

"Our budget," I interrupted.

She nodded and made a face. "There's too much sludge to excavate," she said. "But bioremediation might reach the same result, provided we find the proper electron donor chemical."

Promila explained that, to bioremediate contaminated land in situ, we could introduce naturally occurring organisms such as fungi, yeast, bacteria, and the like. The organisms would execute their natural aerobic functions and consume a food source, while simultaneously consuming any resident contaminants.

"You're saying that microorganisms can eat the chromium?" I asked.

"Yes," Promila said. "This is done all the time in the West."

She told me about the Avco Lycoming Superfund site in Williamsport, Pennsylvania. Since 1929, this industrial park has hosted several manufacturing firms that have produced everything from sewing machines and sandpaper to bicycles and silk plants. Most famously, perhaps, the site was home to Avco Lycoming, a major U.S. manufacturer of jet engines. Over the years, poor housekeeping practices led to the campus's contamination. Industrial wastes were disposed of improperly in a dry well and a coolant well. Metal plating areas dumped or spilled their byproducts inappropriately. Toxic sludge in a holding lagoon leaked out, poisoning local soil and groundwater tables.

In the mid-to-late-1980s, state and federal officials ordered that the facilities be tested. The results showed cadmium present in large quantities, as well as trichloroethylene, a potentially deadly industrial solvent. The site also contained vinyl chloride gas and dichloroethene, a compound used in a slew of industrial applications that has shown moderate oral toxicity in rats.

"And hex," Promila said, adding to the list. "Groundwater tables at the Avco Lycoming site were laced with chromium 6."

"Lovely," I said. "So what did they do specifically?"

Promila told me that, beginning in 1997, the EPA had initiated an intervention at Avco to reductively dechlorinate the aliphatic hydrocarbons.

"Right," I said. "Which in English means what?"

She grinned. "They drastically lowered the cadmium and chromium concentrations. Well, technically, the chromium remains, but the process they chose reduced it from hex to trivalent chromium."

"Which isn't a problem," I said. "Chromium 3 can be dealt with. Okay. So how did they do it?"

"Simple," Promila replied. "With molasses."

"Excuse me?" I said.

She explained that EPA engineers had injected molasses directly into the affected site. The microbes and bacteria already resident in the soil feasted on the molasses. In so doing, they also consumed the oxygen present in the chromium 6, which reverted it to chromium 3.

"Ingenious," I said, stunned.

"Yes," said Promila. "Thanks to the boom in their food supply, the microbes proliferated quickly, which, in turn, broke the toxins down even faster."

"What about byproduct?" I asked.

She shook her head. "The whole process takes place without creating additional deleterious effects to the environment. Once the microbes have done their job, they die, and that is that."

"Poor microbes," I said. "It sounds quite simple."

"It is," she said. "And cost-efficient. Injecting molasses into the ground will not take up much space, nor will it require much energy. And molasses can be obtained for very little money."

Our science staff reviewed our plan and declared it fundamentally sound. We lined up the relevant players and arranged to drill a clutch of wells at Noraiakheda to be used as experiment controls. Next, we mixed up a batch of the molasses material, which we injected into the soil through one of the wells. And that was that. The microbes and bacteria did the rest.

At the time, we had no idea that we were the first group in India's history to attempt a program of this kind. That revelation astonished us, but no more so than the results we obtained from our first experiment. By the end of a series of five readings, the levels of hexavalent chromium in the donor well had dropped dramatically. In fact, they became undetectable. We were also pleasantly surprised to see the levels of trivalent chromium drop.

"How did that happen?" I asked.

Our experts explained that chromium 3 has a natural tendency to bond with rock, so that's what it did. The trivalent chromium had literally fused itself right into the geological substrata, which basically rendered it impotent.

"All this from sugar water," I said.

Promila nodded and laughed.

Our work at Noraiakheda went so well that the project is now under consideration for large-scale implementation. The Central Pollution Control Board would like to apply it to more pockets of contaminated groundwater throughout the Kanpur region. But why stop there? When an inexpensive process yields fast, excellent results, my ears perk up. Where applicable, we'd like to see large-scale implementation become mass-scale implementation.

To be clear, however, no intervention of this sort will solve the overall problem. The chromium problems at Kanpur and similar sites throughout India represent merely the tail end of a much longer and even more abusive cycle. In other words, the situation in Kanpur is merely a blip on the radar.

Consider, for instance, that ninety-seven percent of chromite—the raw mineral from which chromium derives—comes from the Sukinda Valley on the country's eastern shore. Sukinda's Jaipur district is home to twelve separate operating mines, including the largest open cast chrome ore mines in the world. Once upon a time, this cracked landscape sustained life. Now it serves as the dumping ground for more than thirty million tons of overburden, the rock left over after ore is removed from a mineral deposit.

Unfortunately, these deposits overlap the banks of the local Damsala River. Prone to flooding, the Damsala has become the de facto effluent discharge for the region's factories. More than half a million local tribesmen depend on the Damsala for fresh water. They drink from the river, bathe in it, and use it to irrigate crops and water cattle. But hexavalent chromium has infiltrated the water supply in large quantities. Working through the water, the chromium 6 has infiltrated local fish, livestock, produce, milk, and on and on.

In 1995, independent scientists funded by the Norwegian government were amazed to discover that eighty-five percent of deaths in the Sukinda's mining areas and eighty-six percent of deaths in the valley's industrial villages occurred due to chromium 6-related diseases. In villages at distances less than a kilometer from the mining sites, nearly twenty-five percent of residents suffered from pollution-induced disease.

"The pollution and health hazards related to hexavalent chromium are acute," states the scientists' report. It goes on to describe how these health hazards cause "irreparable loss to human health."

We confronted the local pollution control board. In a letter of response, the board stated, essentially, that its hands were tied: "[The problem] is unique, it is gigantic, and it is beyond the means and purview of [the board] to solve the problem."

Clearly, we have our work cut out for us.

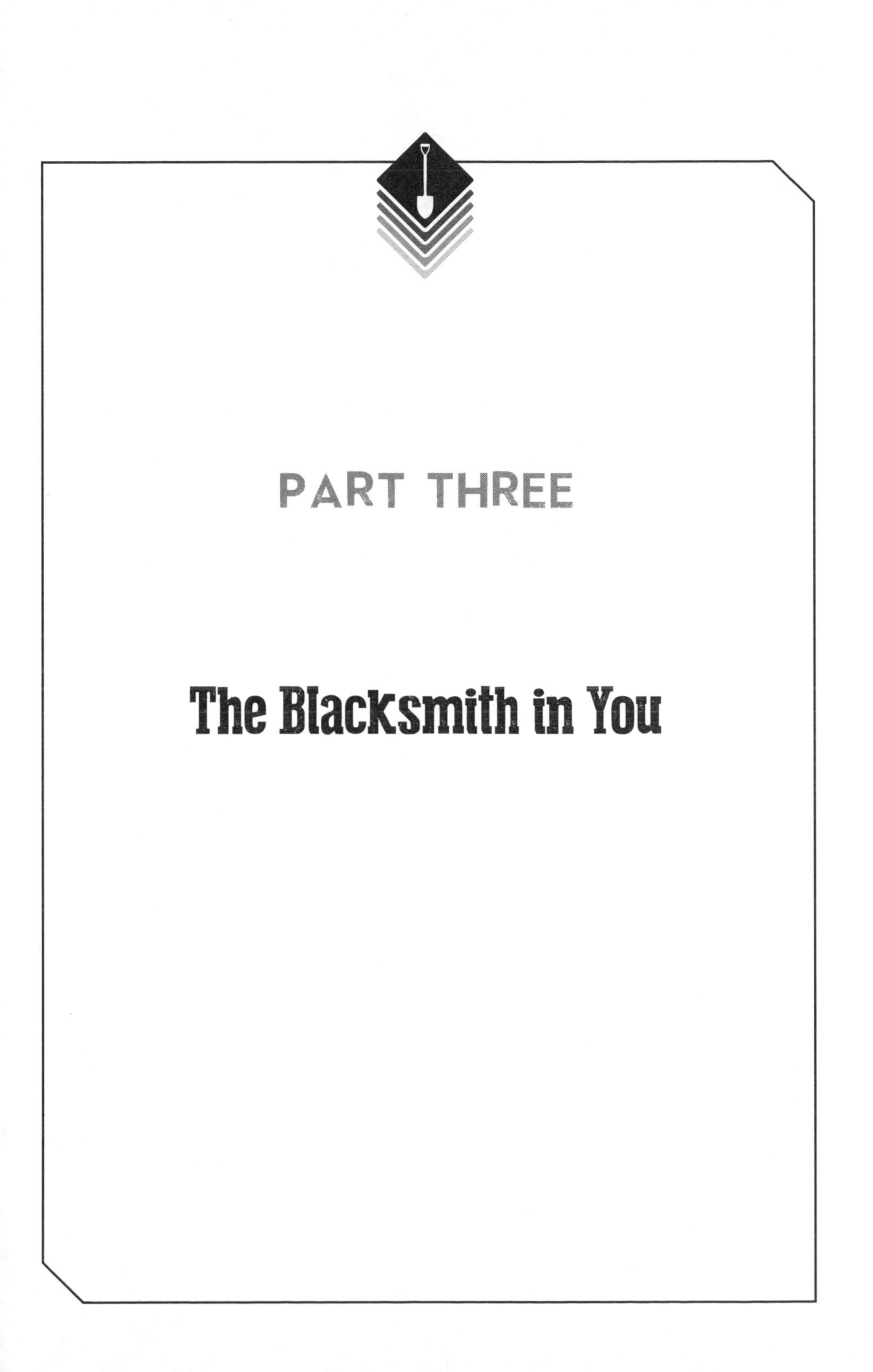

PART THREE

The Blacksmith in You

CHAPTER FIFTEEN

Spread a Message, Change the World

Okay, so that's the bad news. It's true that the top ten pollution problems threatening planet earth pose daunting, and in some cases seemingly insurmountable obstacles. But they are fixable. Really. These problems can be solved. Moreover, when that happens, the biggest single source of death on the planet—let alone an enormous drain on human and environmental well-being, not to mention economic growth—will be vanquished.

I believe this can happen within our lifetimes. How can I be so sure? Because we've already solved many of these problems in the West in about the same period.

In the '50s and '60s, we had air quality in New York and Pittsburg to rival that found in Norilsk and Beijing. We had toxic waste sites at Love Canal to rival those at Kanpur and Huaxi, lead poisoning as bad as Kabwe's in Bunker Hill, Idaho, and sanitation issues everywhere. But all of these problems were solved. Don't get me wrong; we still have pollution problems in the richer countries on earth. But they are smaller than those of their developing nation counterparts and, perhaps most importantly, they're being worked on.

The truth is, creating a pure earth is a problem for realists. It's a task for people who make things happen, rather than just talking environmental problems to death. A realist maps out the scope of the problem, formulates a plan, and gets to work. That means moving soil, installing new technologies, and dealing with communities that need help. A lot of it boils down to following a playbook: Doing what we already know how to do, and have done in the past. Prioritizing cost-effective solutions. Enabling countries to develop resources and structures that will work.

At the end of the day, pollution is really no different from polio: it's a disease that we have the tools to cure. We just need to disseminate the proper treatment.

The most vital component to that dissemination is marshaling the collective will of the world's population. And what's the best way to do that? By turning to the Blacksmith in each of us. Letting everyone be part of the solution.

So what kind of contribution can you make?

Wait! Don't say it. I can already feel you cringing. You think I'm going to ask you to write a big check or do something crazy, like participate in a flash mob. Go on a walkathon. Maybe a danceathon (remember those?). Maybe you think I'll ask you to donate your time, one weekend a month, to march door to door in your voting district, asking your neighbors to sign a petition.

Those things might work. Who knows? But I can offer you something simpler.

I'd like you to *spread a message.*

In today's technological society, there are few things simpler or more powerful than a message. A message is easy to generate, not to mention a renewable resource. A message can spread as fast as the news—and these days, that means pretty damn fast.

A good message has the power to engage a recipient's heart and mind. I'm not talking about some sugary slogan ginned up by Madison Avenue ad agencies. Those mantras, so catchy and clever, convince obese people to consume fast food or people with lung cancer to dole out cigarettes to children.

I'm talking about what I like to call No-Brainer Messages—hear them once, and you're able to digest a complex, vigorous ideation in seconds. From there, you're empowered to do what you like, but perhaps the most powerful action you'll take is also the simplest: spreading this message to others.

Spreading a message is the only way to build its influence. If one person understands a message, ho-hum. If two people get it, they have something in common, something they can talk about; the air between them begins to crackle with a life of its own. If three people get the message, a viewpoint evolves. If four people get it, this viewpoint begins to take on an almost physical mass which, in turn, instigates gravitational pull.

As the fifth, sixth, and seventh people join the core group, the message's effect begins to multiply exponentially. The zeitgeist starts to change in ways that are subtle, yet not to be underestimated. Pressure builds until, eventually, critical mass is obtained. At that point, almost anything can happen.

For a few years now, I've been hugely interested in the sociological phenomenon that writer Malcolm Gladwell covered in his book, *The Tipping Point*. Gladwell's work explored a scenario that we have all seen play out hundreds, possibly thousands of times before anyone thought to put a name on it. That scenario is this: when enough people understand a message, it tends to break across whole cultures like a tidal wave, washing away all that is old and useless and leaving behind a fresh, new landscape of thought on which new generations can build stronger foundations with greater economy, forward thinking, and results.

So what are the messages I'd like you to consider spreading about the brown problem?

Message #1:
Pollution is the biggest killer in the world.

Pollution kills one person for every seven that die. It causes more death and disease then malaria, AIDS, and tuberculosis combined.

This puts things into perspective rather quickly. If you think about it for a moment, I'm sure you will be staggered (as I always am) by the extent of the damage that brown issues cause. This does *not* mean we should take money away from solving those diseases and give them to solving pollution. Of course not. These are terrible problems that deserve all the energy and resources we can give them.

In 2012, the World Health Organization reported that 625,000 people had died from malaria, 1.5 million had died from HIV/AIDS, and 930,000 had died from tuberculosis that year.

That's a total of 3.1 million people, give or take.

The number of people killed by pollution in 2012 was 8.9 million, nearly triple that amount.

With that kind of impact, isn't it time we paid some attention to the brown agenda?

MESSAGE #2:
The victims of pollution are almost all in low- and middle-income countries.

Of the 8.9 million people killed by pollution in 2012, an astonishing 8.4 million of them resided in low- and middle-income countries. Read: they're not in the U.S. or Europe.

There's a simple reason for this. The overwhelming majority of toxic air, soil, and water on earth is found in poorer nations.

The U.S. and Europe have (for the most part) done a good job of protecting our children and our environment. There are a few exceptions where I think the U.S. and Europe could be doing a better job. We have air quality problems in many cities, for instance. Coal-fired power plants and other emissions cause many deaths. We are also uncertain about a range of chemicals that may be causing serious diseases, particularly in children. But make no mistake: the severity of the problems found in the developing world are orders of magnitude worse than those we find in the richer countries.

Exposing people to statistics, photos, and stories about those who live in these toxic sites is, to say the least, very instructive.

Message #3:
Children are hurt more than adults.

This is an important point. Even if you don't have children of your own or even *like* children, you probably understand how vulnerable they are. You probably understand that today's children are the generation that will inherit tomorrow's world and, in many ways, steward it for us in our dotage. We have a vested interest in protecting children because they are the keys to our future.

Message #4:
Big Fortune 500 Companies are not to blame. Local and small enterprises cause almost all of the world's pollution problems.

I know I've said this before, but it bears repeating. Large Western corporations are (generally) not responsible for causing pollution. Rather, quite often they bring best practices to middle-income countries where they set up. With certain rare exceptions, blaming corporate America will not solve pollution problems. Instead, we need to focus our efforts on working with smaller, local companies, which include operators like battery recyclers and gold miners.

What these practitioners really lack isn't willpower, but education and training. The education required to help them clean up their acts need not cost much, and the resulting benefits can equate to huge strides for the world overall, moving forward. Keep in mind that most of these small operators have no idea that they're poisoning their children's heritage. Proper education and training in the use of more effective, toxic-free technologies is the best way to meet their immediate needs while securing a better future for them and us.

MESSAGE #5:
Pollution is *not* inevitable for developing countries.

This is an old mode of thinking that needs to be knocked on the head until it evolves. Many people will claim that the growth required to bring poor countries out of poverty can only come about through actions that cause pollution—that the inevitable cost of growth is pollution, which can be dealt with once the country is richer. After all, such thinkers say, this is what happened when the West industrialized, right?

In today's world, however, this argument is mostly bunk. It's true that old technologies were very polluting throughout the 1930s and '40s, and the West paid the price for them. But there is no reason for new economies to follow that path. In fact, doing so represents a much less inefficient growth trajectory.

The technologies we now enjoy in almost every industry have leapfrogged over those old dirty dinosaurs; no one invests in mercury-based technologies anymore. The newer ones are cheaper to implement and run, more profitable, and inherently cleaner. We no longer build car engines that require leaded gasoline any more than we steer our ships at sea using sextants or dead reckoning. And while it's certainly true that we still suffer many old boneyard industries overseas that are terribly toxic and need to be cleaned up, by and large economic growth relies on efficiency. Meaning that new investments will introduce and leverage more efficient technologies.

As a general rule, remember this: high efficiency and low pollution go hand in hand.

Pollution hinders growth; it is not a natural outcome of it.

We need to keep that front and center as we set our expectations for how current and future industries should conduct themselves.

Message #6:
We can make a big difference in our own world by helping the people in developing nations.

Many pollution problems in other nations affect us directly. For example, air pollution from China reaches across the Pacific to pollute the western coast of the United States, and mercury from artisanal small-scale gold miners contaminates the fish we eventually eat. More subtly, however (though no less powerfully), sharing a pollution-free future builds extraordinary goodwill and gratitude with communities and countries overseas. This, in turn, means a safer world for us all.

Countries that are inhibited from developing their children because of pollution and disease are unable to move out of poverty. Disease and poverty breed the dissention that is a root cause of much of the turmoil in the world. The obvious choice is not always there to make, but helping communities build a strong, healthy environment brings prosperity which, in turn, breaks down the boundaries of tribalism and insularity that lead to disenfranchisement and a lack of contribution to the greater good of our species.

So, there you have it. It may look to you like I'm tilting at windmills. After all, taken in total, the top ten pollutants represent an enormous problem. Yet I have this quiet confidence that we can fix them. Maybe it is the engineer in me, but when I close my eyes to the politics, each of these pollution issues seems like it has a simple solution.

For instance, change the type of diesel fuel that vehicles are allowed to burn.

Introduce a new cook stove to a needy community.

Design and build a treatment plant.

Dig up toxic soil and contain it in proper landfills.

And so on.

You see? It isn't rocket science. Just methodical work, executed one project at a time. And for each project, it is necessary to find the funding, identify the proper technology, and put together the right team that can implement the work successfully.

We take it one country at a time. Each country is a little bit different, with a different set of pollutants. The trick is to focus on the most important problems first—those that kill the most kids. Next, to make sure money is spent on solutions, not just studies. Then, a little bit of support here, a push of encouragement there, good input from an expert or two, and the real change starts to begin. A few pilot projects prove the success of the overall vision; they prove that lives can be saved. More projects follow, funded from country coffers, and soon a new economy is created around anti-pollution and cleanup. The issue becomes embedded in the country's plans. The international response becomes that of a cheerleader, not a driver.

Already, we can see this process at play in so many countries: Mexico, China, India, Indonesia, Brazil . . . the list goes on and on. Needs vary, but the process is the same. It takes many steps over many, many years, but this is the pathway to success.

There's still so much work to do, and you can be a part of it.

So, help us out. Spread the word.

After all, we're all in this together.

FOR MORE INFORMATION

In 2012, by partnering with the World Bank, the European Commission, the Asian Development Bank, and many others, Pure Earth formed the Global Alliance on Health and Pollution.

The GAHP's member agencies include affected low- and middle-income governments, as well as UN agencies and the global development banks. Tasked with implementing solutions to the pollution agenda, the GAHP functions as a kind of clearinghouse, directing monies for development assistance to the direst problems in the most cost-effective way possible. Importantly, GAHP also makes sure that all projects undertaken receive the best personnel available, who leverage the appropriate technical skills to conduct their interventions.

For more information on the GAHP, please visit www.gahp.net.

For more resources and information on how you can contribute to Pure Earth's projects, please visit our home websites: www.pureearth.org and www.blacksmithinstitute.org.

ACKNOWLEDGMENTS

Over the years, it's been my good fortune to have received much incredible support from many incredible people. To call myself grateful here would be an understatement and a disservice to the contributions these people have made toward developing my worldview. The following celebrates just a few of the key players I've come to rely on for various reasons.

First off, I am indebted to the many wonderful people on my staff at Pure Earth. This roster includes professionals of every stripe, from all over the world. To list a few who have been central to the development of this work (with apologies to so many I have missed):

Bret Ericson, Rachael Vinyard, David Hanrahan, Meredith Block, Corinne Ahearn, Rohan Lawrence, Kira Traore, Angela Bernhardt, Marlo Mendoza, Promila Sharma, Budi Susilorini, Jenny Sunga, Mag Sim, Sarita Gupta, Bill Landesman, Sara-Kate Gillingham, Katherine Mechner, Sun Xuebing, Vladimir Kutnetzov, Sandra Gualtero, Wang Leyan, Russell Dowling, Sandy Page-Cook (to whom I am also most lucky to be married, and who runs our *Journal of Health and Pollution*), Drew McCartor, Lina Hernandez, Daniel Estrada, Jack Caravanos, Jen Marraccino, Lilian Corra, Larah Ibanez, Petr Sharov, Johnny Ponce, Julius Ngalim, Fatou Sow, Babu Sengupta, Steve Laico, Duong Thi To, Rovshan Abbasov, Sergei Sharapenko, Peter Hosking (Oz!), and Amalia Aborde.

Although it might be unusual to acknowledge institutions, there are a few that have seen the wisdom in dealing with pollution, and have been early adopters for the cause, despite it being complicated or contentious. The European Commission is one of them, an organization that really understands chemicals, pollution, and the extent to which they can injure human beings and the environment. The World Bank has also been a passionate advocate for the poisoned poor. The Asian Development Bank has also held the banner. To these institutions—and those just beginning to support our mission—I say thanks so much! The chance to make a difference is enormous, and the case for action is all too clear.

For our non-institutional financial support, for helping on our board of directors, for donations of every kind, and for the invaluable service of acting as a lightning rod for ideas, I thank: Conrad Meyer (an awesome chairman!), Sheldon Kasowitz (equally wonderful as our prior chair), Ron Reede, Ken Rivlin, Paul Brooke, Josh Ginsberg, Josh Mailman, David Mechner, Alex Papachristou, Paul Roux, Asif Sheik, Colin Stewart, Charlotte Treifus, Mark Machin, Ian King, Sid Sandilya, Gille and Maria Concordel, Ira Riklis, David Wichs, Sam and Scott Zinobar, Ruben Kraiem, and Charles Goodyear.

Many wonderful technical experts have donated their time and expertise to our work. For their priceless contributions, I thank: John Keith, Anne Reiderer, Dr. Philip Landrigan, Nadia Gilladorni, David Hunter, Leona Samson, Rock Brynner, Frances Beinecke, Paul Dolan, Nick Albergo, Pat Breysse, Tim Brutus, Jim Darling, Denny Dobbin, Bruce Forrest, Ian von Lindern, Margaret von Braun, Steve Gorman, David Green, Bill Drayton, Pascal Haelfliger, Joe Hayes, Gil Jackson, Eric Johnson, Barbara Jones, Don Jones, Mukesh Khare, Ira May, Terry Oda, Jerry Paulson, Anne Riederer, Dave Richards, Stephan Robinson, Brian Wilson, and Adriana Damionava.

No progress could be made abroad without aid and counsel from our friends in governments and institutions the world over. This list includes: Senator Pia Cayetano, Neric Acosta, Ramon Paje, Jill Hanna (the most wonderful and passionate advocate for the poisoned poor in the development community—go Jill!), Yves Prevost, Jostein Nygard, Mauricio Limon, Naomi Chakwin, Kristalina Georgieva, Heinz Leuenberger,

Yumkella Kandeh, Mathy Stanislaus, Karen Mathiasen, Nathalie Gysi, Stephan Robinson, Tim Kasten, Linda Greer, Achim Steiner, Jacob Duer, Tisha Chaterjee, Mary Barton-Dock, Warren Evans, Rachael Kyte, Helena Naber, Jairam Ramesh, Chris Sheldon, Patricia Moser, Bindu Lohani, Pierre Quiblier, Bart Edes, Amy Leung, Enrique Ona, and Cristiana Pasca Palmer.

Over all these years, the team at Great Forest has been wonderful. In capacities all too often unsung, its members continue to link the many strands of sustainability and international work with our clients and friends. Not insignificantly, Great Forest has backstopped me for the last decade, enabling me to work on Pure Earth without worrying about supporting my family. They have been there for me throughout, and mostly while I have been absent. Thanks guys, I owe you all! You are the greatest! Thanks to Amy Marpman, Anna Dengler, Ross Guberman, Joe Matos, Maya Shenkman, Sheila Sweeney, David Troust, Tom Cstari, Barbara Fonseca, Nefertiti Ruff, Caroline Hazarian, Nate Holmes, Beth Kimball, Kelly Kent, Ken Richards, Gabriel Montano, Erwin Brisso, Kevin McNab, Sandra Robishaw, Todd Sutton, Joe Romuno, Yovelice Collado, Mayline Johnson, and many former colleagues and associates.

Other friends who have had the patience to be there at the right time include Gaston Silva, Rock Brynner, David Hunter, Leona Samson, Jim Doran, Peter Hosking, Pat Cassidy, Indira Sandilya, Jake Lindsay, Jeff Elmer, Scott Salmirs, Greg Dinella, Tom Felderman, Ralph Scopo, Jay and Marie Wholley, Bill and Maggie McLeod, Silda Wall, and Erik Simon.

I am blessed with a wonderful partner, Sandy, and three kids who fill the days with pollution-free pleasure—Max, Milo, and Alice. Love ya, guys.

And of course, thanks go to Martha Kaplan, my wonderful literary agent; Jeffrey Goldman at Santa Monica Press, a delightful publisher; and to Damon DiMarco, my co-author. While the story you've read has been mine, make no mistake—Damon is the reason it might be interesting. He is a wordsmith par excellence.

And finally, to one special person: Karti Sandilya, with whom I have the enormous pleasure of spending a great deal of time each year traveling the world on behalf of Pure Earth. Karti's network is amazing, but it is his gentle sense of purpose and wisdom that makes everything seem possible. We should all be so lucky to have a friend and advisor like him. Thank you, Karti!

PHOTO CREDITS

All uncredited photographs were provided by Richard Fuller and Pure Earth.

Samit Car, Occupational Safety and Health Organisation of Jharkhand: xxiv (top left)

Dr. Jack Caravanos: vii (top, bottom right); xx (top left)

Bret Ericson: xv (bottom)

Andreas Haberman: xvii (top)

Peter Hosking: cover; ii (middle); iv–v

Rakesh K. Jaiswal, Eco Friends: xxiii (all)

Stanislav Lvovsky: xviii–xix

Andrew McCartor: ix (all); x (top); xi (all); xx (top right, bottom)

Tom Murphy: xii (bottom)

Babashov Nicat: xv (top)

Nelson Pampolina: xiv (top)

ABOUT THE AUTHORS

Richard Fuller is the founder and president of Pure Earth (formerly the Blacksmith Institute). Considered the world's leading expert on toxic issues, Pure Earth works closely with governments, local stakeholders, and international organizations around the world to combat the proliferation of toxic pollution in forty-five countries. In 2012, Fuller also led the formation of the Global Alliance on Health and Pollution, a collaborative body tasked with coordinating resources and activities to clean up chemicals, waste, and toxic pollution in low- and middle-income countries. The GAHP currently has thirty-two members, including many affected countries, UN agencies, and bilateral and multilateral donors such as the World Bank.

Damon DiMarco is the author of the oral histories *Tower Stories: An Oral History of 9/11* and *Heart of War: Soldiers' Voices from the Front Lines in Iraq.* With Baiqiao Tang, he wrote *My Two Chinas: The Memoir of a Chinese Counter-Revolutionary,* which won the endorsement of His Holiness, the Dalai Lama. A classically trained actor, DiMarco also co wrote *The Actor's Art and Craft* and *The Actor's Guide to Creating a Character* with William Esper. For more information, please visit www.damondimarco.com.

Bryan Walsh (Foreword) is the foreign editor and senior writer on energy and the environment for *Time* magazine.

For more information about Pure Earth,
please visit www.pureearth.org.